REASSESSMENT OF METALS CRITERIA FOR AQUATIC LIFE PROTECTION

Priorities for Research and Implementation

Edited by

Harold L. Bergman
University of Wyoming

Elaine J. Dorward-King
Kennecott Utah Copper Corporation

Proceedings of the Pellston Workshop on
Reassessment of Metals Criteria for Aquatic Life Protection
10–14 February 1996
Pensacola, Florida

SETAC Technical Publications Series

Publication sponsored by the Society of Environmental Toxicology
and Chemistry (SETAC) and the SETAC Foundation for Environmental Education

Published by SETAC Press
Pensacola, Florida

Cover by Michael Kenney Graphic Design and Advertising
Copyediting and typesetting by Wordsmiths Unlimited

Library of Congress Cataloging-in-Publication Data

Pellston Workshop on Reassessment of Metals Criteria for Aquatic Life Protection (1996 : Pensacola, Fla.)
   Reassessment of metals criteria for aquatic life protection : proceedings of the Pellston Workshop on
Reassessment of Metals Criteria for Aquatic Life Protection, 10–14 February 1996, Pensacola, Florida / edited by
Harold L. Bergman, Elaine J. Dorward-King.
      p. cm. -- (SETAC technical publications series)
   Includes bibliographical references.
   ISBN 1-880611-04-x (alk. paper)
1. Metals--Environmental aspects--Congresses. 2. Metals--Toxicology--Congresses. 3. Aquatic organisms--Effect of
metals on--Congresses. 4. Metals--Environmental aspects--Research--Congresses. 5. Metals--Toxicology--Research--
Congresses. 6. Aquatic organisms--Effect of metals on--Research--Congresses.
I. Bergman, Harold L. II. Dorward-King, Elaine J. (Elaine Jay), 1957- III. Title. IV. Series.
QH545.M45P44 1996
577.6'2753--dc21   97-3646
                                                                                                              CIP

© 1997 Society of Environmental Toxicology and Chemistry (SETAC)
Published by SETAC Press
SETAC Press is an imprint of the Society of Environmental Toxicology and Chemistry.
No claim is made to original U.S. Government works.

International Standard Book Number 1-880611-04-x
Printed in the United States of America
04 03 02 01 00 99 98 97   10 9 8 7 6 5 4 3 2 1

♾ The paper used in this publication meets the minimum requirements of the American National Standard for
   Information Sciences—Permanence of Paper for Printed Library Materials, ANSI Z39.48-1984.

Reference Listing: Bergman HL, Dorward-King EJ, editors. 1997. Reassessment of metals criteria for aquatic life
protection: priorities for research and implementation. SETAC Pellston Workshop on Reassessment of Metals
Criteria for Aquatic Life Protection; 1996 Feb 10–14; Pensacola FL. Pensacola FL: SETAC Pr. 114 p.

The SETAC Technical Publications Series

The SETAC Technical Publications Series was established by the Society of Environmental Toxicology and Chemistry (SETAC) to provide in-depth reviews and critical appraisals on scientific subjects relevant to understanding the impacts of chemicals and technology on the environment. The series consists of single- and multiple-authored or edited books on topics reviewed and recommended by the SETAC Board of Directors and approved by the Publications Advisory Council for their importance, timeliness, and contribution to multidisciplinary approaches to solving environmental problems. The diversity and breadth of subjects covered in the series reflects the wide range of disciplines encompassed by environmental toxicology, environmental chemistry, and hazard and risk assessment. Despite this diversity, the goals of these volumes are similar; they are to present the reader with authoritative coverage of the literature, as well as paradigms, methodologies and controversies, research needs, and new developments specific to the featured topics.

The SETAC Technical Publications are useful to environmental scientists in research, research management, chemical manufacturing, regulation, and education, as well as to students considering or preparing for careers in these areas. The series provides information for keeping abreast of recent developments in familiar subject areas and for rapid introduction to principles and approaches in new subject areas.

Reassessment of Metals Criteria for Aquatic Life Protection: Priorities for Research and Implementation presents the collected papers stemming from a SETAC-sponsored Pellston Workshop held in Pensacola, Florida, 10–14 February 1996. The workshop focused on unresolved scientific issues and needed significant research in the area of current and possible future approaches to regulating metals in aquatic environments. This book was written at the workshop, revised through reviews by the participants, and completed after the workshop. The book represents the views of the authors and the participants in attendance. It has not been peer reviewed in the classic sense of having anonymous reviewers critique it, but it has been reviewed by each workshop participant and extensively scrutinized for technical and editorial consistency and correctness by the workshop Steering Committee and a paid editorial consultant.

Preface

This book presents the proceedings of the 26th Pellston Workshop, held 10–14 February 1996 in Pensacola, Florida. Like previous workshops, participation was limited to invited experts from government, academia, and industry who were selected because of their experience with the workshop topic. The workshop provided a structured environment for the exchange of ideas and debate such that consensus positions would be derived and documented for some of the issues surrounding the science and regulatory practice related to water quality criteria for metals. The proceedings reflect the current state-of-the-art of these topics and focus on an assessment of 1) regulatory practices, 2) environmental toxicology, 3) environmental chemistry, and 4) fate and transport modeling, all related to metals in aquatic environments.

Acknowledgments

The Pellston Workshop on Reassessment of Metals Criteria for Aquatic Life and publication of the workshop proceedings were made possible through the financial support of ARCO, Cominico, Eastman Kodak Company, International Council on Metals and the Environment, International Lead Zinc Research Organization, Merck Manufacturing Division, Mining Association of Canada, National Mining Association, and Pacific Gas & Electric Company. The content of this publication does not necessarily reflect the position or the policy of any of these organizations or the United States government, and no official endorsement should be inferred.

The workshop organizers gratefully acknowledge the participants for their enthusiastic commitment to the workshop's objectives and for their cooperative efforts in the preparation of this book. We are also deeply indebted to the SETAC/SETAC Foundation Office staff and volunteers who contributed to the success of the workshop. Special thanks go to Rod Parrish, Greg Schiefer, Linda Longsworth, and Leslie Long for their hard work in planning and implementing the workshop and to Lisa Credeur for her diligence in preparing the final manuscript. We also greatly appreciate Mimi Meredith's valuable assistance in editing the manuscript throughout the review process.

Errata Sheet

REASSESSMENT OF METALS CRITERIA FOR
AQUATIC LIFE PROTECTION:
Priorities for Research and Implementation

The following organizations should also have been acknowledged for their financial support of the workshop: the Electric Power Research Institute, the International Copper Association, and the Utility Water Act Group.

Foreword

This workshop was a continuation of a series of successful workshops called the "Pellston Workshop Series." Since 1977, twenty-six workshops have been held at Pellston and several other locations to evaluate current and prospective environmental issues. Each has focused on a relevant environmental topic, and the proceedings of each have been published as a peer-reviewed or informal report. These documents have been widely distributed and are valued by environmental scientists, engineers, regulators, and managers because of their technical basis and their comprehensive, state-of-the-science reviews. The workshops in the Pellston Series are as follows:

- *Estimating the Hazard of Chemical Substances to Aquatic Life.* Held in Pellston, Michigan, June 13–17, 1977. Proceedings published by the American Society for Testing and Materials, STP 657, in 1978.

- *Analyzing the Hazard Evaluation Process.* Held in Waterville Valley, New Hampshire, August 14–18, 1978. Proceedings published by The American Fisheries Society in 1979.

- *Biotransformation and Fate of Chemicals in the Aquatic Environment.* Held in Pellston, Michigan, August 14–18, 1979. Proceedings published by The American Society of Microbiology in 1980.

- *Modeling the Fate of Chemicals in the Aquatic Environment.* Held in Pellston, Michigan, August 16–21, 1981. Proceedings published by Ann Arbor Science in 1982.

- *Environmental Hazard Assessment of Effluents.* Held in Cody, Wyoming, August 23–27, 1982. Proceedings published in a SETAC Special Publication by Pergamon Press in 1986.

- *Fate and Effects of Sediment-Bound Chemicals in Aquatic Systems.* Held in Florissant, Colorado, August 11–18, 1984. Proceedings published in a SETAC Special Publication by Pergamon Press in 1987.

- *Research Priorities in Environmental Risk Assessment.* Held in Breckenridge, Colorado, August 16-21, 1987. Proceedings published by SETAC in 1987.

- *Biomarkers: Biochemical, Physiological, and Histological Markers of Anthropogenic Stress.* Held in Keystone, Colorado, July 23–28, 1989. Proceedings published in a SETAC Special Publication by Lewis Publishers in 1992.

- *Population Ecology and Wildlife Toxicology of Agricultural Pesticide Use: A Modeling Initiative for Avian Species.* Held in Kiawah Island, South Carolina, July 22–27, 1990. Proceedings published in a SETAC Special Publication by Lewis Publishers in 1993.

- *A Technical Framework for [Product] Life-Cycle Assessments.* Held in Smuggler's Notch, Vermont, August 18-23, 1990. Proceedings published by SETAC in January 1991, with second printing in September 1991 and third printing in March 1994.

Workshop Participants and Contributing Authors

William J. Adams
Kennecott Utah Copper Corporation
Magna, UT

Herbert E. Allen[a]
University of Delaware
Newark, DE

Gerald T. Ankley
USEPA
Duluth, MN

Larry W. Ausley
North Carolina Division of
  Environmental Management
Raleigh, NC

Harold L. Bergman[a]
University of Wyoming
Laramie, WY

Finn Bro-Rasmussen
Technical University of Denmark
Lyngby
DENMARK

C. Robert Cappel
Eastman Kodak Company
Rochester, NY

John P. Connolly
HydroQual, Inc.
Mahwah, NJ

William Davison
Lancaster University
Lancaster
UNITED KINGDOM

Nico de Rooij
Delft Hydraulics
Delft
NETHERLANDS

Charles G. Delos
USEPA
Washington, DC

Donald R. Di Bona
Medical University of South Carolina
Charleston, SC

Miriam L. Diamond
University of Toronto
Toronto, ON
CANADA

Dominic M. Di Toro[a]
HydroQual, Inc.
Mahwah, NJ

Elaine J. Dorward-King[a]
Kennecott Utah Copper Corporation
Magna, UT

Russell J. Erickson[a]
USEPA
Duluth, MN

Guy Ethier
International Council on Metals
  and the Environment
Ottawa, ON
CANADA

Kathy L. Godtfredsen
EVS Consultants
Seattle, WA

David J. Hansen
USEPA
Narragansett, RI

Russell S. Kinerson, Jr.
USEPA
Washington, DC

James R. Kramer
McMaster University
Hamilton, ON
CANADA

Norman E. LeBlanc
Hampton Roads Sanitation District
Virginia Beach, VA

Christopher M. Lee
International Copper Association
New York, NY

Samuel N. Luoma
U. S. Geological Survey
Menlo Park, CA

Jack S. Mattice[a]
Electric Power Research Institute
Palo Alto, CA

Joseph S. Meyer
University of Wyoming
Laramie, WY

P. Rodney Parrish
SETAC/SETAC Foundation
Pensacola, FL

James F. Pendergast
USEPA
Washington, DC

E. Michael Perdue
Georgia Institute of Technology
Atlanta, GA

Richard C. Playle
Wilfrid Laurier University
Waterloo, ON
CANADA

Donald B. Porcella
Electric Power Research Institute
Palo Alto, CA

Mary C. Reiley[a,b]
USEPA
Washington, DC

William L. Richardson
USEPA
Grosse Ile, MI

Gregory E. Schiefer
SETAC/SETAC Foundation
Pensacola, FL

Christian E. Schlekat
University of South Carolina
Columbia, SC

Jerald L. Schnoor
University of Iowa
Iowa City, IA

James F. Stine
Baltimore Gas & Electric Company
Baltimore, MD

William A. Stubblefield
ENSR Consulting & Engineering
Fort Collins, CO

Edward Tipping
Institute of Freshwater Ecology
Cumbria
UNITED KINGDOM

Dik van de Meent
RIVM
Bilthoven
NETHERLANDS

Alexander Viteri, Jr.
State of Alaska
Juneau, AK

John C. Westall
Oregon State University
Corvallis, OR

Christopher M. Wood
McMaster University
Hamilton, ON
CANADA

a) Workshop Steering Committee Member
b) Unable to attend workshop

Contents

Harold L. Bergman, Elaine J. Dorward-King, Herbert E. Allen, Dominic Di Toro,
Russell J. Erickson, Jack S. Mattice, Mary C. Reiley

James F. Pendergast, Larry W. Ausley, Finn Bro-Rasmussen, C. Robert Cappel,
Charles Delos, Elaine J. Dorward-King, Guy Ethier, David J. Hansen,
Norman E. LeBlanc, Christopher M. Lee, Alexander Viteri Jr.

Chapter 6: Environmental fate and transport 71

Jerald L. Schnoor, John P. Connolly, Dominic M. Di Toro, Nico de Rooij, Miriam Diamond, Russell S. Kinerson Jr., Donald B. Porcella, William L. Richardson, James F. Stine

Appendix: Discussion-Initiation Papers 81

List of Tables

List of Figures

Acronyms and Initialisms

AVS	acid volatile sulfide
AWQC	ambient water quality criteria
BAT	best available technology
BOD	biological oxygen demand
CCC	criterion continuous concentration
CCME	Canadian Council of Ministers of the Environment
CCREM	Canadian Council of Resource and Environment Ministries
CMC	criterion maximum concentration
CWA	Clean Water Act
DIC	dissolved inorganic carbon
DOC	dissolved organic carbon
DOM	dissolved organic material
EDTA	ethylenediamine tetraacetic acid
EU	European Union
FA	fulvic acid
FWPC	Federal Water Pollution Control Act
GESAMP	Group of Experts on Scientific Aspects of Marine Pollution
GIS	Geographic Information System
ICME	International Council on Metals and the Environment
IFCS	Intergovernmental Forum on Chemical Safety
IMO	International Maritime Organization
IPPC	Integrated Pollution Prevention Control
LTI	Limno-Tech, Inc.
NEDSS	National Environmental Decision Support System
NMPS	National Modeling Prototype Studies
NPDES	National Pollutant Discharge Elimination System
NRDC	National Resource Defense Council
OECD	Organisation for Economic Co-operation and Development

PD	probabilistic dilution
POM	particulate organic matter
POTW	publicly owned treatment works
SD	static dilution
SQC	sediment quality criteria
SRSS	spatially resolved steady state
SRTV	spatially resolved time variable
SSIM	sparingly soluble inorganic material
TDS	total dissolved solid
TIE	toxicity identification evaluation
TSCA	Toxic Substances Control Act
TMDL	total maximum daily load
TSS	total suspended solids
UNCED	United Nations Conference on Environment and Development
U.S.	United States
USEPA	United States Environmental Protection Agency
WER	water-effect ratio
WET	whole effluent toxicity
WHAM	Windermere Humic Aqueous Model
WLA	wasteload allocation
WQC	water quality criteria
WQO	water quality objectives
WQS	water quality standards

Chapter 1

Executive summary

The SETAC Workshop, "Reassessment of Metals Criteria for Aquatic Life Protection: Priorities for Research and Implementation," was held to reassess the scientific and engineering basis for the metals criteria and standards. Both regulatory practice and recent scientific advances were reviewed, and recommendations were made. Discussions during the workshop suggested a consensus in the scientific and regulatory communities that a departure is needed from the purely observational and empirical practice of evaluating the biological impact of metal exposure on organisms. Organism response is now being more mechanistically linked to details of accumulation in a target tissue and to physiological responses to this accumulation. Mathematical models of both processes can be the integrating vehicles. Linking these models to receiving-stream exposure models is the next step toward a more fundamental approach to regulatory decisions. These models will explicitly take into account what has previously been termed _bioavailability_ and have the potential to be predictive across a wide range of environmental conditions. It is the opinion of the workshop attendees that this direction marks a turning point in the development of a workable methodology for assessing the effects of metal exposure on organisms. Besides endorsing the development of a mechanism-based approach for assessing metal effects, several short-term pragmatic recommendations are provided for use while the mechanistic models are developed.

Four workgroups focused discussions for each of the represented disciplines or interest areas: 1) regulatory practices, 2) environmental toxicology of metals, 3) chemical speciation and metal toxicity in surface fresh waters, and 4) environmental fate and transport. The following is a brief synopsis of the findings from each workgroup (please see the full descriptions of these recommendations in Chapters 3 through 6).

Regulatory Practices Workgroup

The Regulatory Practices Workgroup affirmed support for the recommendations of the January 1993 Annapolis meeting (see Chapter 2). Further, the workgroup recognized that any advances in science and regulatory practice need to be communicated to those who make regulatory decisions. For any regulatory approach to succeed, people involved in regulatory and permitting decisions need to be educated in the relevant portions of the scientific and engineering disciplines represented by the workgroups. The relevant knowledge cannot be reduced to cookbook applications of well-understood principles. Regrettably, the current trend is that government support for technical assistance and training is declining. It is strongly recommended that this trend be reversed so that more scientifically sound and economically efficient solutions to regulatory problems can be reached.

Other recommendations from the Regulatory Practices Workgroup's discussions include the following:

- Allow regulatory authorities the flexibility to use either appropriately sensitive toxicity tests (*e.g.*, whole effluent toxicity [WET]) or numerical limits on metals, or both, when setting standards or making permitting decisions. Given the current knowledge that site-specific factors can significantly modify the toxicity of a metal, the reliance on only numerical limits to determine when a metal quality criterion is exceeded may not always be justified.

- Adopt a tiered method using both chemical and biological tests to decide whether to regulate an effluent. A tiered approach that starts with simple screening and later uses increasingly more comprehensive and costly information was felt to offer a better way to make regulatory decisions in a sound and cost-effective manner. The workgroup recommended a tiered approach that 1) starts with an evaluation of the potential to exceed a water quality standard based on data using clean chemistry and the dilution/loading model used by the permitting authority; 2) continues with a site-specific adjustment to the ambient water quality criteria (AWQC) and model calculations based on an empirical dissolved organic carbon (DOC) partitioning model or a verified mechanistic model, if available; 3) continues with laboratory toxicity tests with the effluent of concern, using the most sensitive organisms from the original AWQC for the metal under consideration; and 4) ends with an abbreviated toxicity identification evaluation (TIE) using a metal chelator.

- Quantify the actual cumulative safety factor in the regulatory decision process, and make it apparent to all.

- Develop a toxicity and chemical speciation model, validate it with multiple datasets, and use it to replace the current criteria. This model, or suite of similar models, will go to the heart of the toxicity problem by addressing the effects of metals at the sites of action associated with toxicity, while incorporating the influences of site-specific water chemistry. As a result, these models would make the new AWQC directly applicable to individual waters without relying on the current approach of using site-specific studies.

Environmental Toxicology Workgroup

The Environmental Toxicology Workgroup supports the recent policy allowing the use of dissolved metal instead of total recoverable metal values for permits, when appropriate, as a move in the right direction for improving regulatory practice by incorporating new scientific understanding. However, the workgroup emphasizes that the current criteria approaches, even with this modification, are not based on a sound understanding of why metal toxicity varies with water quality. The workgroup provides a number of short-term and longer-term recommendations, all oriented toward developing the information needed for incorporation of mechanistic information about metal toxicity into water quality criteria and standards:

- Use the water-effect ratio (WER) based on dissolved concentrations, when appropriate, as a reasonable interim solution in water quality regulations until it can be replaced by regulations that have a mechanistic basis. In the short term, the current use of the WER should be evaluated, the procedure refined as necessary, and water regulation professionals educated in its application.

- Explore replacing the existing AWQC approach with a gill modeling or tissue residue approach that is mechanistically based. Supporting research should be undertaken for development of the gill modeling approach; information is needed regarding studies of 1) gill residue versus mortality for copper, cadmium, nickel, silver, lead, zinc, and chromium; 2) multiple organisms; 3) metal mixtures; 4) different water quality conditions; 5) pulse exposure effects and time course; 6) applicability to longer-term exposures; 7) mode of action, emphasizing clarification of mechanisms such as transport of calcium or sodium ions and tissue accumulation of those ions; and 8) kinetics.

- Collect and evaluate existing datasets relating toxicity to key water quality characteristics influencing bioavailability, such as pH, hardness, and DOC. These empirical, toxicity-based relationships can be used to help validate a gill-binding/metal toxicity model and support current evaluations of AWQC compliance.

- Continue to employ acute-to-chronic ratios for setting standards that are protective of chronic endpoints until the mechanisms underlying chronic toxicity and the linkages between tissue residues and chronic toxicity are understood.

- Use existing datasets and develop new data to assess the relationships among acute toxicity of whole-body residues, organ-specific residues, or tissue/organ biochemical fractionations for key species of fish and invertebrates.

- Research is needed to elucidate mechanisms behind toxicity in invertebrates and the linkage to metal residues; this should include the possible effects of metals on sodium and calcium ion transport, as has been demonstrated in fish.

- In the next generation of criteria for protecting aquatic ecosystems, consider multiple routes of exposures (*e.g.*, dietary as well as dissolved). Sediments and water should not be considered separately.

- For chronic toxicity, establishing linkages between specific tissue residues (*e.g.*, gill or liver residues in fish) and mechanisms of toxicity should be a high priority in the development of the tissue-residue, mechanism-based approach to regulation.

- Consider ecosystem response in evaluating chronic toxicity. It is essential to identify which processes and effects on individuals determine how populations and communities respond.

Environmental Chemistry Workgroup

The Environmental Chemistry Workgroup considered how current knowledge about metal speciation in aquatic systems could be used in improving the mechanistic understanding of aquatic metal toxicity, and in assessing and modeling that toxicity. The workgroup provides a number of recommendations regarding modeling and measuring chemical speciation, focused on better descriptions of the relationship between metal speciation and biological effects:

- Use of models such as MINTEQA2 and MINEQL$^+$ would be improved by inclusion of the Humic-ion Binding Model V. Use of these modified models would enable a wide range of users to carry out metal speciation calculations, and the models should be incorporated in any proposed changes in approach for metals regulation.

- The Windermere Humic Aqueous Model (WHAM) and other models incorporating Model V should be used to evaluate the potential for correlation of toxicity with specific aqueous metal species, given total and dissolved metal concentration. Examination of past datasets containing total and dissolved metal concentrations and toxicity data would show if adverse effects are discernible. Emphasis on instances where toxicity was/was not exhibited relative to the metal species present would assist in evaluating regulatory policy.

- A fundamental investigation is needed to relate pH and dissolved metal to suspended solids on the basis of measurable characteristics such as particulate organic matter (POM) and metal oxyhydroxides. Meanwhile, to provide the link between dissolved metal concentration and total recoverable metal needed for transport and permitting purposes, the empirical K_d approach is recommended if the system is adequately characterized with respect to pH, total suspended solids (TSS), hardness, salinity, and flow ranges.

- At the present, determination of dissolved metals concentration is uncertain, and it is not clear that the potential problems associated with filtration techniques can be resolved; other techniques that measure dissolved species directly should be assessed (*e.g.*, voltametric or gel techniques). In addition, standardized methods used to determine metals data should be developed; these should routinely include the use of "clean" or "ultraclean" procedures.

- WHAM, modified MINTEQA2, and modified MINEQL$^+$ models should be used comprehensively in the design of experiments and in the interpretation of data with respect to toxicology research. Use of these models will permit integration of metal speciation and the proposed gill response model.

- Collation of information and research on forms of metals in food web components are needed for modeling metal accumulation and effects.

- The term *bioavailability* is often misused in the context of equilibrium speciation. The key point deals with correlation of metal species to toxicity.

- As work on the mechanistic gill model progresses, the following points should be borne in mind to ensure a useful product: 1) extensive sets of data covering the ranges of concentrations for system variables need to be developed to relate solution composition to toxicity, 2) the model should be consistent with all sets of data used in the development, and 3) transition from the present "proof-of-principle" phase to definitive models needs to be made expeditiously.

Environmental Fate and Transport Workgroup

The Environmental Fate and Transport Workgroup encourages the more consistent application of models that adequately 1) predict the chemistry of metals in water bodies, including their transformations between free ion and complexed forms; 2) predict the partitioning of metals between water column, sediments, and aquatic organisms; and 3) consider the sources of metals loading to the ecosystem. This will involve greater sophistication than the simple dilution equation that is commonly used for setting permit limits. The workgroup acknowledges that the best model is the simplest one that is appropriate for the task. The workgroup developed the following recommendations:

- Adopt a probabilistic modeling approach for riverine systems rather than applying static dilution (SD) methods.
- Design models to protect both the water column and sediment.
- Continue the use of the translator methodology for permits that use either direct measurement or regression equations to calculate the ratio between total and dissolved metal fractions.
- Adopt watershed approaches using a Geographic Information System (GIS) to manage multiple discharges and nonpoint source loadings (near and far field). The model should treat metals, nutrients, organics, and chemicals that exist as multiple, interconverting species.
- Include time dependency functions in spatially resolved models to predict sediment quality, diurnal variations, storm events, and resuspension.
- Account for multiple routes of chemical exposure to biota (*e.g.*, through inclusion of a food web model).
- Test models under a variety of field conditions (long- and short-term, near and far field).
- Develop a National Environmental Decision Support System (NEDSS) to support modeling efforts. Similarly, National Modeling Prototype Studies (NMPS) should be initiated to validate fate models used in permitting.

Introduction

Harold L. Bergman, Elaine J. Dorward-King, Herbert E. Allen,
Dominic Di Toro, Russell J. Erickson, Jack S. Mattice, Mary C. Reiley

Water quality regulations to protect aquatic life from exposure to metals and other chemicals have been in place in the United States (U.S.), Canada, and the European Union (EU) for several decades. Concentrations of metals are regulated in part on the basis of water quality criteria (WQC)[1] or water quality objectives (WQO)[2] that are derived from the results of acute and chronic toxicity tests with aquatic organisms exposed to the metals in laboratory tests. However, metal toxicity in ambient site waters can differ from toxicity in laboratory waters due to variation in water quality characteristics such as temperature, pH, hardness, alkalinity, suspended solids, or dissolved organic carbon. Water quality criteria or water quality objectives generally do not account for these variations in water quality characteristics, except in some cases for water hardness, based on comparisons of toxicity among laboratory waters that differ in hardness and other correlated water quality characteristics. Moreover, in implementing the ambient WQC or WQO for many of the metals, most regulatory agencies require that "total recoverable" or "total" analytical determinations be used to assess compliance with water quality standards (WQS)[1], although it is acknowledged that such "criteria may be overly protective" (USEPA 1985a). In the absence of reliable analytical methods and models for better specifying toxicity expected at a site, the USEPA (1984, 1994) developed guidelines for modifying WQC based on toxicity tests in site water.

Debate over appropriate metals criteria for protection of aquatic life

During the past several years, the debate about the appropriateness of existing metals criteria or objectives and analytical methods for implementing metal standards has continued and intensified among the regulated community, regulators, and interested scientists. Many of the broad scientific issues underlying this debate were addressed in August 1992 at a SETAC Pellston Workshop and in the resulting book, *Bioavailability: Physical, Chemical and Biological Interactions* (Hamelink *et al.* 1994).

During recent years, several agencies have also sponsored workshops to review these issues. In a December 1990 USEPA workshop entitled "Recommendations for Revising National Water Quality Criteria Guidelines," several speakers suggested that metals criteria and standards were overly stringent, and that some measure of the "bioavailable" or

---

[1] *Water quality criteria* (WQC) is a term used in the U.S. Clean Water Act (CWA) to represent a scientific assessment of ecological effects caused by a particular pollutant; WQC to protect aquatic life are established by the United States Environmental Protection Agency (USEPA) using a specified procedure (USEPA 1985b) that requires a minimum dataset on the toxicity of the pollutant to aquatic biota. In the U.S., these WQC, along with other information related to local conditions, are used by states to develop *water quality standards* (WQS).

[2] The term *water quality objective* (WQO) is used in the EU and is essentially analogous to WQC as used in the U.S., but the uses and derivation procedures for WQO are more diverse and country-specific (See Chapter 3).

"toxic" fraction of metals in surface waters was both necessary and achievable. Discussion of these issues continued at a January 1993 USEPA workshop in Annapolis, Maryland, on "Aquatic Life Criteria for Metals" (USEPA 1993). More recently, several workshops have addressed hazard identification and risk assessment of metals in the Organisation for Economic Co-operation and Development (OECD) framework (*e.g.*, OECD 1995).

A group of international scientific and regulatory experts in attendance at the USEPA Annapolis Workshop offered a series of short-term and long-term recommendations to resolve continuing issues with regulation of metals. The short-term recommendations addressed were these:

1) Needed usage of clean chemistry techniques

2) Continued usage of "total recoverable" metal determinations for mass balance calculations and permit limitations

3) Adoption of "dissolved metal" determinations to better approximate the bioavailable fraction of metals when evaluating AWQC and standards

4) Continued usage of WERs as a strategy for addressing site-specific toxicity of metals

5) Listing of under- and overprotective factors and assumptions that underlie metals criteria and standards for use by discharge permit writers

6) Adoption of special provisions for regulating organometallic compounds such as metalized dyes (USEPA 1993)

The USEPA has initiated changes in its regulatory practice for metals related to all but the last two of these Annapolis recommendations. A particularly important outcome of these recommendations is that the USEPA Office of Water has begun to recommend that "dissolved metal" be used to determine compliance with WQS (Prothro 1993; USEPA 1993).

For long-term recommendations, participants in the 1993 Annapolis Workshop listed a series of important research needs that should be addressed to support improved, future regulatory approaches for metals. These recommendations encompassed research needs in environmental toxicology, chemistry, and fate of metals, as well as regulatory implementation. Several of the most important recommendations on research needs included: 1) tests of hypotheses concerning "free metal ion activity" as the best measurement of the toxic, bioavailable fraction of aqueous metals; and 2) the relative importance of waterborne versus dietary routes of metal exposure in aquatic organisms. The participants also recommended that researchers and research funding agencies focus on metals that pose the greatest regulatory problems and on those that offer the greatest opportunity for addressing the most critical research needs. Further, the participants recommended that a workshop be convened to review existing knowledge about metals in the aquatic environment, list research priorities, and develop critical research plans with the goal of providing needed scientific bases for the next generation of regulations to protect aquatic life from exposure to metals. The SETAC Metals Workshop was organized to address this last recommendation.

Purpose and objectives of the SETAC Metals Workshop

The overall purpose of the SETAC Metals Workshop was to assist the research, regulated, and regulatory communities by recommending the best approaches for assessing the toxicity of metals in natural surface waters and by identifying research needed to resolve continuing issues regarding such assessments. The most important, long-term, practical benefits anticipated from this workshop are the continued movement toward more scientifically defensible WQC for metals and, in so doing, the formation of the basis for the next generation of aquatic life criteria and regulations for metals.

The specific objectives for the SETAC Metals Workshop were as follows:

- Review the current state-of-knowledge about metals in the aquatic environment, including environmental toxicology, chemistry, and fate as related both to fundamental knowledge and to current and future regulatory strategies for metals.

- Identify gaps in fundamental knowledge about the environmental toxicology, chemistry, and fate of metals; list critical research that would fill these gaps in fundamental knowledge; and develop a plan that would help researchers and funding agencies identify priorities for research on metals.

- Assess the consequences of current knowledge, and achievable advances in knowledge about metals, on improvements in regulatory practice.

- Publish a workshop report for rapid dissemination to the interested research, regulated, and regulatory communities.

Organization and agenda for the SETAC Metals Workshop

The workshop was held 10–14 February 1996 in Pensacola, Florida, with 44 invited, international participants from government, business, and academia. The participants represented a diversity of experience related to the scientific and regulatory issues addressed at the Metals Workshop, and many were present at earlier workshops and meetings in North America and Europe, where the debate outlined in a preceding section, "Debate over appropriate metals criteria for protection of aquatic life," was joined. Prior to the workshop, participants were assigned, based on a distribution of expertise and perspective, to one of four workgroups: 1) regulatory practice for metals, 2) environmental toxicology of metals, 3) chemical speciation and metal toxicity in surface fresh waters, and 4) environmental fate and transport.

The first full day of the workshop was devoted to a series of presentations by invited experts on 1) regulatory practices for metals, from regulatory agency, industrial, municipal, and international perspectives; 2) environmental toxicology of metals; 3) chemical speciation and metal toxicity in surface fresh waters; 4) environmental fate and transport; and 5) an initial synthesis of perspectives and potential alternatives for resolving the scientific and regulatory issues about metals in the aquatic environment. The extended abstracts of these presentations are included as appendixes to this book.

The following two days were filled with separate workgroup discussions and writing sessions, along with brief plenary session reports and discussions on progress of each of the workgroups. A number of difficult, crosscutting issues were also debated among representatives of the different workgroups during breaks and specially arranged sessions among members of groups to facilitate information exchange.

The consensus final reports, conclusions, and recommendations from each of the four workgroups were then presented and discussed in plenary session on the fourth and last morning of the workshop, followed by a presentation and discussion of the overall, integrated workshop summary and conclusions.

Following the workshop, the participants, working through the workgroup chairs and steering committee members and with the assistance of SETAC staff, reviewed and revised all chapters of the workshop report. The final workgroup reports are presented in Chapters 3 through 6 of this book, with the workgroup chairs and participants listed as authors of each chapter.

The positions, conclusions, and recommendations presented in this book about WQC for protection of aquatic life represent the general consensus achieved by the participants attending the SETAC Metals Workshop. The book has been reviewed by Workshop participants but has not been subjected to outside peer review by qualified experts who were not in attendance at the Workshop.

References

[CWA] Clean Water Act. 33 U.S.C. §1251 *et seq.* (June 30, 1948). Also titled Federal Water Pollution Control Act.

Hamelink JL, Landrum PF, Bergman HL, Benson WH. 1994. Bioavailability: physical, chemical and biological interactions. Boca Raton FL: Lewis. 239 p.

[OECD] Organisation for Economic Co-operation and Development. 1995. Proceedings of the OECD Workshop on aquatic toxicity testing of sparingly soluble metals, inorganic metal compounds and minerals. 1995 Sep 5–8; Ottawa, ON, Canada. Ottawa, ON: International Council on Metals and the Environment (ICME). 92 p. (draft)

Prothro MG. 1993. Office of water policy and technical guidance on interpretation and implementation of aquatic metals criteria. Memorandum from Acting Assistant Administrator for Water. Washington DC: USEPA Office of Water. 7 p. Attachments 41 p.

[USEPA] U.S. Environmental Protection Agency. 1984. Guidelines for deriving numerical aquatic site-specific water quality criteria for modifying national criteria. Duluth MN: Environmental Research Laboratory. EPA/600/3-84/099 or PB85-121101.

[USEPA] U.S. Environmental Protection Agency. 1985a. Ambient water quality criteria for copper - 1984. Washington DC: Office of Water. EPA/440/5-84/031. PB85-227023. 142 p.

[USEPA] U.S. Environmental Protection Agency. 1985b. Guidelines for deriving numerical national water quality criteria for the protection of aquatic organisms and their uses. Springfield VA: National Technical Information Service (NTIS). PB85-227049.

[USEPA] U.S. Environmental Protection Agency. 1993. Water quality criteria: aquatic life criteria for metals. *Federal Register* 58(108):32131–32133.

[USEPA] U.S. Environmental Protection Agency. 1994. Interim guidance on determination and use of water-effect ratios for metals. Washington DC: Office of Water, Office of Science and Technology. EPA/823/B-94/001. 154 p.

Chapter 3

Regulatory practice for metals

James F. Pendergast, chair
Larry W. Ausley, Finn Bro-Rasmussen, C. Robert Cappel,
Charles Delos, Elaine J. Dorward-King, Guy Ethier, David J. Hansen,
Norman E. LeBlanc, Christopher M. Lee, Alexander Viteri Jr.

Historically, the uncontrolled release of metals into the aquatic environment has caused adverse effects on some aquatic systems. Recognizing that the release of metals is an inevitable result of natural and anthropogenic activities, efforts have been made to determine environmental concentrations (WQS or WQO) which, if met, would not jeopardize the integrity of biological communities in surface waters. Numerous physical, chemical, and biological interactions need to be understood to accurately establish these standards or objectives and to translate them into appropriate control strategies. Initial WQC were developed by the USEPA that were recognized as being overly protective in waters that had greater potential to bind metals than the dilution waters used in the toxicity tests on which the criteria were based. The initial criteria developed are regarded as reflecting worst-case situations. These initial criteria and subsequent standards did not consider environmental interactions. It was assumed that all forms of metals were as bioavailable and toxic in ambient water as the metals present in the toxicity tests forming the basis of the criteria. The potential lack of technologically feasible solutions and/or extreme societal costs to meet these standards has driven the need to reexamine the basis for these standards and improve them.

In parallel with the efforts to establish WQS for metals, existing engineering pollution control activities were advanced, resulting in substantial reductions in metal releases from anthropogenic sources and reduced metal concentrations in the water column and sediments. There have also been numerous improvements in monitoring techniques along with an increased understanding of metal interactions. Developments in biocriteria have increased the ability to measure the status of the receiving ecosystem (USEPA 1990). In North America, the advent of WET procedures with sensitive aquatic organisms and the concomitant TIEs suggest that the implementation of pretreatment technology standards has largely eliminated effluent metal toxicity from municipal treatment plants (Grothe _et al._ 1996). For example, statewide application of WET testing to about 500 municipal and industrial discharges in North Carolina, in concert with TIEs, indicated that about two percent had metals-related toxicity problems (Ausley, personal communication).

Even with these improvements, it is not always possible to meet all numerical metal standards. Municipal facilities are still having difficulty meeting permit limits based on criteria for protection of aquatic life, even when such criteria are corrected for site-specific factors affecting toxicity. The improved understanding of the behavior of metals in the environment has provided the impetus to reexamine the criteria and their use for regulatory purposes. To this end, the Regulatory Practices Workgroup examined the current regulatory practices for control of metals, identified barriers to regulatory reform, and provided recommendations for better control of metals based on current scientific understanding.

Current regulatory practices

Brief summaries follow, reviewing the current regulatory structures for regulating metals by establishing water quality criteria and permitting metals discharges.

United States

In the U.S., the CWA establishes the regulatory framework that controls the discharge of pollutants, including metals, into the aquatic environment. The CWA identifies federal and state responsibilities to define and achieve acceptable water quality levels for metals. USEPA is required to develop WQC for the protection of aquatic organisms through the use of the best available scientific information and to update the criteria as new information becomes available. The USEPA uses data from laboratory studies to develop criteria that are intended to protect most of the nation's aquatic species most of the time, even during rare events, such as low stream flow or high discharge flow. The criteria concentration is calculated to estimate the LC50[1] of the theoretical ninety-fifth percentile genus in the available database.

The CWA requires states to develop WQS by assigning established criteria to protected water uses. To do this, states modify the national USEPA criteria to reflect the species and the receiving water characteristics within the state. Most states, however, do not have the resources to modify or develop the criteria in a scientifically defensible manner and have adopted the USEPA criteria as their standards. Finally, states convert these ambient standards into National Pollutant Discharge Elimination System (NPDES) permit effluent limits for point source discharges to surface waters that meet the state standards. Those permit limits are generally established using a state-defined worst-case situation, which usually consists of a defined low-effluent dilution and a maximum facility design flow. The CWA allows the states to decide how to conduct this computation. Several unquantified and unspecified safety factors have been built into both the standard-setting and permitting processes.

The permit limits established by this system are considered absolute values that incur liability when exceeded. Permit exceedances are violations of both federal and state laws and are enforceable by either government or by private citizens.

Canada

The Canadian framework for national environmental quality guideline development and implementation is described by Gaudet *et al.* (1995). As is summarized in that paper, Environment Canada, in collaboration with provincial and territorial ministries in the Canadian Council of Ministers of the Environment (CCME), develop nationally consistent guidelines for environmental quality in Canada (Canadian Council of Resource and

---

[1] LC50 is the concentration lethal to 50% of the test organisms in an acute toxicity test (exposure periods ranging from 24 to 96 hours).

Environment Ministries [CCREM] 1987). These guidelines for water quality, sediment quality, tissue-residue quality, and soil quality define clear, scientifically defensible, and nationally accepted indicators of environmental quality for protecting, sustaining, and restoring aquatic and terrestrial ecosystems and their uses in Canada. They are developed using nationally approved scientific protocols (CCME 1991, 1993, 1996a, 1996b, 1996c) to ensure consistency, transparency, and scientific defensibility in the process.

An initial emphasis on the development of nationally consistent guidelines for aquatic ecosystems led to the publication of the first edition of the Canadian Water Quality Guidelines (CCREM 1987) for major Canadian uses of water. More recently, the Canadian guidelines have been expanded to encompass other important components of aquatic and terrestrial ecosystems, including marine water quality; water quality for agricultural uses (irrigation and livestock water); marine and freshwater sediment quality; tissue-residue quality for protection of aquatic life and wildlife consumers; and soil quality guidelines for agricultural, residential/park land, commercial, and industrial land uses.

There are three fundamental guiding principles in the development and application of the Canadian environmental quality guidelines. First, they embody a national goal for environmental quality of no observable adverse effect on aquatic and terrestrial ecosystems over the long term. Second, Canadian environmental quality guidelines are developed for major uses of land and water in Canada. This resource-use–based system provides a socioeconomically relevant framework for environmental management that flexibly accommodates a broad range of sites or issues including complex, multi-use scenarios. Finally, although Canadian environmental quality guidelines provide a nationally consistent scientific basis for management decisions such as the development of substance-, site-, or issue-specific objectives or standards, they do not directly incorporate management considerations nor are they intended to serve directly as management objectives without due consideration of such factors. National guidelines are recommendations that are based on the most current scientific information. Objectives, in contrast, are established or agreed upon by the affected parties and may reflect site-specific scientific information as well as societal values, costs, and technological limitations.

Environmental quality objectives are set by most provinces and territories, taking into account the unique social, economic, and biophysical characteristics of their jurisdictions. Such objectives are implemented through provincial legislation and policy and may be adopted as legally enforceable standards. Objectives may also be set on an ecosystem, regional, or site-specific basis, taking into account the above characteristics. Objectives may cross several jurisdictional boundaries, and public and stakeholder participation and societal values for the system could play a strong role in influencing the final objectives.

European Union

In the European Union, industrial chemicals, including metal and metal compounds, are regulated under total or minimum harmonization rules, the minimum rules permitting member countries to establish and enforce stricter rules. The hazard identification/assessment processes that form the scientific basis for setting WQO are generally coordinated through the fully harmonized classification and labeling requirements of Directive 67/548 (Seventh Amendment of 1992). Water quality objectives in Europe are analogous, not identical, to WQS in the U.S. The actual setting, however, of WQO as EU standards has been required only for List I chemicals, *i.e.,* the list of 129 chemicals in Directive 76/464 (1976). Among the metals, this list includes cadmium, cadmium compounds, and mercurials. Numerically, WQO are established on the basis of the evaluation of ecotoxicity to aquatic life, bioaccumulation, and biodegradation and by taking into consideration zero-effect levels. Whereas the WQO for List I chemicals are considered as community standards, the national governments have an obligation to establish WQO for List II chemicals, which include most of the ecotoxicologically relevant metal compounds.

National and local governments (counties and sometimes municipalities) have the authority to establish discharge limit values. The discharge of chemicals for which a WQO has been established must meet the WQO outside the defined mixing zone. In countries where specific receiving water planning is instituted, a further regulation-defined dilution zone may be allowed in accordance with the local authority's rules for water quality.

According to a new directive recently passed by the EU Council of Ministers, guidelines for the Integrated Pollution Prevention Control (IPPC) have been established that require discharge and emission reductions from industrial point sources according to the principle of best available technology (BAT). Discharge (or emission) limit values will have to be set by member states under the provisions of this new directive, while quality standards already set in Directive 76/464 (and daughter directives), including those for metal and metal compounds, generally will continue to apply as community standards. Voluntary schemes for environmental management and certification are made acceptable (and encouraged) as tools, and they also regulate discharges to the aquatic environment.

Risk assessment

The current international approaches to environmental management of metals and metal compounds, including establishment of criteria and standards, rely to a large extent on some form of risk assessment. Numerous assumptions are used and are built into the risk assessment process. There are several different applications of risk assessment, including the following:

- Classifying as "dangerous for the environment" for the purposes of packaging and labeling of products containing metals or their compounds, or for transport of bulk metals, metal compounds, or metal-containing products, such as in the EU, International Maritime Organization/Group of Experts on Scientific Aspects of Marine Pollution (IMO/GESAMP), or OECD
- Assessing environmental risk for the purpose of pre-manufacture or pre-marketing notifications of metal-containing products, as in the U.S. Toxic Substances Control Act (TSCA), the EU Seventh Amendment to Directive 67/548 (1992), or the Canadian pre-manufacturing notification

- Controlling metals levels in wastewater discharges, surface waters, and soils, such as in the U.S. under the CWA
- Reassessing metals or metal compounds under existing chemicals regulations and guidance as in the EU, Canada, and OECD

All of the preceding examples of risk management applications lead to environmental management decisions that implicitly or explicitly integrate risk assessment into criteria setting. The intent in risk assessment is to accommodate the inherent uncertainty that results from variability in the temporal or spatial distribution of metals in the environment, coupled with variability in environmental conditions and organism responses. This accommodation is extremely important for metals where the "implementation of metals criteria is complex due to the site-specific nature of metals toxicity" (Prothro 1993). For example, if regulatory decisions on metals are based on the total recoverable form and ignore site-specific conditions, the decisions may be either overprotective, imposing an unfair burden on dischargers, or underprotective, leading to an unacceptable environmental risk. Additionally, international differences in the establishment of WQS for metals and in regulations restricting discharge may erect unintended barriers to trade and commerce.

Current regulatory practices and barriers to environmental management of metals

Specific examples of problems associated with current regulatory practice, primarily in the U.S., are provided in this section. These problems are recognized by many practitioners as being barriers to protective and cost-effective environmental management.

Use of dissolved metals for measurement in water quality criteria

Dissolved metal is operationally defined as the metals in a solution that pass through a 0.40 μm to 0.45 μm filter. A part of what is measured as dissolved metal is actually soluble hydrated or complexed metal ions combined with particulate metal that is small enough to pass through the filter openings, or metals that are absorbed to or complexed with organic colloids and ligands. Some or all of the small particulate-bound metal and the soluble complexed metal may not be toxic to aquatic organisms. *Total recoverable metal* is defined as the measurement of metals after acid digestion, according to USEPA's promulgated analytical methods (USEPA 1984b). *Particulate metal* is operationally defined as the difference between total recoverable metal and dissolved metal in a sample.

Historically, in the U.S., WQC for metals have been set exclusively in terms of total recoverable metal (USEPA 1986). The USEPA has changed that policy based on recommendations from the 1993 Annapolis workshop (Prothro 1993; USEPA 1993). The new policy recommends use of dissolved metal concentrations to set and measure compliance with WQS. The reason given for the change in the policy was that dissolved metal more closely approximates the toxicity of a metal in the water column than does total recoverable metal. The adoption of this new policy did not change USEPA's view that the existing WQC, based on total recoverable metal, are scientifically defensible. USEPA recommends that states use risk management decisions to consider sediment, food web effects, and other fate-related issues when deciding to use total recoverable or dissolved metals criteria for developing their WQS.

This practice, when used, has resulted in adoption of WQS that better reflect the toxicity of a metal. However, this practice is not uniformly followed in the U.S., nor is it followed in the EU. In addition, the practice has resulted in raising a number of new questions, such as whether the accumulation of metals in sediments causes water quality problems and whether the overall process for calculating a permit limit has become too complicated. Furthermore, although the use of a dissolved metal value is generally considered to be a better indicator of metal toxicity, it is not clear that it is always a better indicator than a total recoverable metal value.

Use of numerical translators to convert dissolved metal to total recoverable metal

In the U.S., effluent limits are generally expressed in terms of the total recoverable metal. In addition, modeling of metals requires consideration of both the dissolved and total recoverable forms of metals. By regulation, NPDES permit limits in the U.S. must, with few exceptions, be expressed as total recoverable metal. USEPA current policy describes methods for converting dissolved fraction values to equivalent total recoverable values (Prothro 1993; USEPA 1996). Given these conversion translator mechanisms, states using dissolved fraction-based criteria are able to calculate an equivalent total recoverable metal value that is appropriate for a site.

Differences between the chemistry of an effluent or wastewater and that of the receiving water are expected to change the distribution of dissolved and absorbed metals in the receiving water. Because U.S. regulations require NPDES discharge limits to be expressed as total recoverable metal, states are required to translate metal concentrations back to total recoverable metal concentrations for permitting purposes. A translation factor ("translator") is used to estimate the concentration of total recoverable metals in the surface water, which equates to the dissolved concentration in that water. The translated metal mechanism is then applied to a receiving water's metal standard to calculate the metal concentration in the total form. This value can then be used to calculate a permit limit using mass balance model techniques.

Translator mechanisms do pose problems in their application. In some cases, information is insufficient to develop a site-specific translator; generic information must then be applied from other sites. This introduces uncertainty into the evaluation, and guidance is available to help permit writers and modelers with these situations (USEPA 1996). Also, the amount of dissolved metal in some ambient waters may be independent of the amount of particulate metal; this tends to occur following higher flow or scouring events (Connolly 1985). As a result, the ability of generic translator mechanisms to reliably predict the dissolved fraction of a total recoverable metal measurement is questionable.

Lack of clean sampling and analytical techniques

Problems can arise for industries and municipalities when regulations and assessments for metals are based on erroneous analytical data. If clean[2] sampling and analysis techniques were not used, then historical databases that contain information on metals concentrations in effluents and ambient waters that are below one part per million may be invalid due to possible contamination. Regulatory decisions should not be made on the basis of this questionable data because contamination errors of one to two orders of magnitude are common, and errors of three to four orders of magnitude have been demonstrated when clean techniques are not used to measure metals below the part-per-million range (Adeloju and Bond 1985; Berman and Yeats 1985; Bloom 1993).

Modern methods for clean or ultraclean collection, handling, filtration, and analysis of samples must be used to collect new effluent and receiving or ambient water samples for metals when the expected concentrations are lower than one part per million, if confidence is to be placed in the analysis of these samples. USEPA (1995a, 1995b) has published guidelines that describe these clean techniques.

The cost of using clean techniques will typically run $2000 to $5000 per sample event per site, depending on the number of metals analyzed, their concentrations, and the number of filtrations required to determine the dissolved fractions (Cappel, personal communication). It is essential to include in each sampling event the appropriate number of sample "blanks" designed to check for possible contamination at each stage.

Some agencies or dischargers may resist the use of clean techniques due to the costs. However, data on metals concentrations below the part-per-million level collected without using clean techniques are essentially useless, and their collection is a waste of time and money.

Water-effect ratio approach

The WER approach to developing site-specific WQC recognizes that characteristics of site water, effluent, and the specific metal can combine to influence the toxicity of the metal to aquatic organisms at a specific site. This toxicity differs from toxicity determined in laboratory water used to derive national WQC. The WER (ratio of the toxicity of the metal in side-by-side tests using site and laboratory water) is the empirical toxicological determination of the difference in metal toxicity, attributed to site water, that is used to numerically adjust the national WQC.

Initial guidance on methodologies for conducting WER determinations (USEPA 1983b) was modified following the January 1993 Annapolis meeting (USEPA 1994b). Two approaches were proposed: 1) use of simulated downstream water (combination of effluent and upstream water) and 2) use of actual site water from large water bodies to which the site-specific criterion was to apply. Metal-specific recommendations of species and tests sensitive at WQC concentrations were intended to measure metal-binding affinities ap-

---

[2] The term *clean technique* refers to sampling and analysis techniques that reduce contamination and enable accurate and precise measurement of trace metals in fresh and salt surface waters. This includes using appropriate quality control and quality assurance procedures and establishing detection limits that are adequate.

propriate to the criterion concentration because WERs and metal-binding increase with decrease in total metal concentration (Allen and Hansen 1996). USEPA guidance recommends the development of WERs based on both total recoverable and dissolved metal, with dissolved metal WERs likely being most consistent and total recoverable metal WERs varying with water quality parameters such as organic carbon and solids concentrations (USEPA 1994b).

In addition to the guidance on determination and use of WERs for metals, there is recent guidance (USEPA 1994b) on performing the recalculation option for deriving site-specific WQC that reflects the sensitivities of families of aquatic biota and their life stages at specific sites. This guidance has been used to update national criteria datasets. This latter approach is preferred because deletion of data for nonresident species further reduces the already limited national databases. Many test species may be toxicological surrogates of untested local species, and national criteria most often will be lowered because of the computational methodology and not because of any real difference in species sensitivities.

The WER studies have been conducted most often for copper. The WER has not been widely needed for other metals. In practice, WERs of dissolved copper in salt water do not differ greatly from unity and are quite stable for a site; however, WERs for total recoverable metals at these sites are more variable. Water-effect ratios of copper in freshwater are highly variable. In general, WERs for publicly owned treatment facilities primarily receiving domestic wastewater indicate that there are few copper toxicity problems.

The WER is generally being used with success, and it is currently the only method for obtaining a reasonable criterion value for copper in the presence of elevated dissolved organic matter. However, there are some problems with using WER, including 1) it is a relatively complex approach that is cumbersome for small dischargers, 2) it has limited value in situations where stable organo-metal complexes are formed, and 3) there are methodological questions regarding the form of metal being added or the equilibration time for the added metal salt.

Predicting sediment toxicity

The developments in our understanding of the toxicity of metals in sediments have also provided tools to help identify whether sediment problems are likely to occur. Anecdotal reports suggest that problems with metals in sediment in many parts of the country are often associated with uncontrolled metals loads discharged by inefficient industrial or mining processes in the late nineteenth century and first half of the twentieth century, but more recent activities are sometimes also implicated (Perwak *et al.* 1980; Hudson-Edwards *et al.* 1996; Spliethoff and Hemond 1996).

Present USEPA policy does not directly regulate discharges that might produce unacceptable metals concentrations in sediments (USEPA 1996a). Exceedance of sediment criteria or safe sediment levels for chemicals appears to occur for the following reasons, as opposed to accumulation from chemical concentrations in water at or below the WQC permitted level: historical events resulting in the discharge of significant quantities of chemicals to surface waters that have exceeded normal fate, degradation, and redistribution processes; nonpoint source inputs and unpermitted discharges to surface waters; and the failure of waste treatment systems that are fully permitted. Exceedances occur even with the best waste treatment designs. Excursions above the design criteria and WQC can have significant impacts on sediment chemical concentrations for long periods of time even though the WQC are met 99% of the time.

In summary, however, it is believed that WQC are protective of organisms that live in sediments. This conclusion is based on basic principles of chemical thermodynamics and equilibrium partitioning. It is also based on the knowledge that the permitted discharge levels to the water column are designed to be, on average, well below the water quality criterion, thereby providing additional protection and capacity for chemical concentrations in the sediment to increase without immediately exceeding toxic levels. Even though simple modeling analyses suggest that WQC are protective of sediments, this has not been fully confirmed by field studies.

In the near term, the "Contaminated Sediments Management Strategy" (USEPA 1994a) (presently under review) will provide an equilibrium partitioning-based numerical sediment quality criteria (SQC) for metals. The strategy specifies the use of SQC based on acid-volatile sulfide partitioning and interstitial water-metal concentration in sediments. Effluent permitting will require the development of water quality models that link water column and sediment metals in space and time. The models must be capable of computing seasonal cycles of metal-binding phases with sediment depth and must provide insight into the link between WQC and SQC. Further, the models must be able to predict metals fate where sediments are disturbed, such as in dredging or sediment remediation.

USEPA policy of independent application of standards

Approaches for assessing water bodies to determine if WQS for metals are achieved inherently incorporate assumptions about the toxicity of chemicals, potential for exposure, and other interrelated environmental variables. It is therefore likely that, when those assumptions do not adequately describe the actual characteristics of the water body, the prediction of adverse effects may be far from accurate. These discrepancies between predicted effects and measured or observed effects can create regulatory dilemmas in setting and enforcing appropriate standards.

One regulatory policy implemented by the USEPA has a significant bearing on the way that criteria are used to establish or enforce standards by USEPA or the states (Davies 1991). This policy, which has become commonly known as "Independent Applicability," includes the following concepts:

1) each of three regulatory water quality monitoring systems (chemical-specific criteria, WET tests, and biological assessments [generally, aquatic macroinvertebrate assessments]) hold equal weight in establishing compliance with WQS;

2) contravention of any of the three measurements results in a violation of WQS; and

3) conflicting results between any two or more of the different measurements cannot be used to overrule the finding of another.

The Independent Applicability policy is greatly undermined by significant technical weakness in any one of its three legs. Advances in the understanding of how metals behave in the environment, both physically and toxicologically, cause practitioners to question the use of the Independent Applicability approach. For example, controversy arises when predictions of toxic impacts of chemicals are not supported by findings of toxicity tests that, with the exception of possible differences in species sensitivity, use the same concentration and effect assumptions as the chemical-specific approach. In contrast, biological criteria may indicate water quality problems although all metal WQC are achieved. Moreover, there are situations where the three measurement systems would not necessarily be expected to agree. An example would be a bioaccumulative metal, like mercury, where an accurate chemical-specific criterion based on bioaccumulation may be exceeded without causing effects measured by WET analyses or changes in aquatic population structure, though additional tissue monitoring may detect problems. This is a much different situation than that for metals like copper and zinc, where the concentrations may exceed the WQS without evidence of toxicity or adverse effects.

Alternatively, a "weight of evidence" approach would allow closer scrutiny of disagreements, investigation of possible causes, and facilitate informed decisions about compliance with environmental standards. An example of implementing such an approach is presented in more detail in a discussion-initiation paper in this document (Ausley and Reid, Appendix). This type of approach is useful in compliance analysis of NPDES discharges, but it is limited to the demonstration of presence or absence of effect and may not be valuable in evaluating sub-effect concentrations that may be a necessary part of total maximum daily load (TMDL) development for toxic substances in a water body.

Calculation of effluent permit limits

Once a wasteload allocation (WLA) has been calculated by a regulatory authority, it is usually used to develop an effluent limit. In the U.S., this is called an "NPDES permit limit." Although the WLA may be expressed in terms of either the dissolved or total recoverable form of a metal, the NPDES permit limit must, by rule, be expressed in the total recoverable form, with only a few exceptions such as where other U.S. effluent regulations pertain to a specific form of a metal (*e.g.*, hexavalent chromium). USEPA promulgated this rule at 40 CFR 122.45 (USEPA 1984a). Consequently, regulatory authorities need to convert between the total recoverable and dissolved metal forms when constructing fate and transport models for calculating permit limits.

This rule was written to ensure that unintended environmental effects will not occur once an effluent is discharged into a surface water. USEPA's concern was that if only the dissolved form of a metal was controlled by a permit limit, then the uncontrolled particulate form of the metal could be discharged in quantities that could become dissolved in the ambient surface water and lead to water quality problems (USEPA 1984a). This concern is alleviated if reliable data show that the dissolved fraction of metal in the effluent is a reliable predictor of the dissolved fraction of metal in the ambient water, if not for all facilities then at least for a few types of facilities (*e.g.*, publicly owned treatment works [POTW]).

This rule does not present an insurmountable barrier to the effective and credible control of metals. However, it does introduce another layer of complexity to the process, thus resulting in greater difficulty in convincing permitting authorities to adopt dissolved metal standards and a greater information burden to permit writers for developing credible permits.

Safety factors to account for uncertainty

The concept of safety factors is a part of environmental risk management that attempts to account for uncertainties by reserving part of an allowable emission or by using a stringent evaluation of the bounds of uncertainty. Safety factors are widely used to compensate for the uncertainties inherent in risk assessment. They are also used in the application of the "precautionary principle" in the EU. The development of a safety factor reduces the amount of a pollutant that a facility can release, and in the case of surface water, it reduces the amount of a pollutant that can be discharged to that water. Safety factors can result from choices of regulators during criteria development (*e.g.*, the use of water containing little DOC or particulates yields additional safety when the criteria are applied to typical ambient water). In WQS, the use of the total recoverable form of the metal to address the unknown fate of metals yields additional safety. In the calculation of TMDLs, safety factors can result from either an explicit reservation of part of the TMDL or by the selection of worst-case values to quantify environmental or facility characteristics. In NPDES permits, safety factors can result from assuming the highest effluent discharge event occurs during a rare low-dilution event, although the probability of both events occurring simultaneously is small.

Ideally, the safety factor will be determined by the decision-maker, or the safety factor will be an explicit part of a risk assessment process and thus is known to the decision-maker. In practice, though, individual safety factors are applied at several interim steps in the decision process, and thus the overall safety factor is not usually quantified or even quantifiable in all applications. As a result, the overall safety factor may be very large, and more importantly, the decision-maker may be unaware of it. The USEPA (1983a), for example, attempted to quantify the cumulative safety factor in the WLA process and found it to be an order of magnitude. The resulting report did not characterize the actual practice of developing WQS and NPDES permits. Although the report cannot be considered as a precise assessment of the overall safety factor associated with NPDES permits based on WLAs, it does provide an estimate of the overall factor.

The lack of information about uncertainties and quantification of the safety factor can present a barrier to effective risk management, risk communication, and credible decision-making. A better approach to current practice may be to calculate a TMDL and permit limit without using individual safety factors, and then apply an overall safety factor at the end. In this way, the overall safety factor will be quantified and known when the regulatory decision is made, thus promoting credible risk management decisions.

Recommendations

First, the Regulatory Practices Workgroup reemphasizes its support for the recommendations of the January 1993 Annapolis meeting (Prothro 1993; USEPA 1993). Most of the recommendations were adopted and implemented by the USEPA, and these provided the impetus for more credible control of metals discharges. Second, the workgroup recognizes that any advances in science and regulatory practice must be communicated to those who make regulatory decisions. Thus, the workgroup strongly recommends that regulatory authorities increase the education and training of federal, regional, provincial, state, and local regulatory staffs regarding the rules, policies, guidances, and science of metal regulation. This is required if scientifically defensible and consistent regulation of metals is to be achieved.

After investigating the unimplemented Annapolis recommendations and considering the practices of metal controls over the last three years, the workgroup developed the additional recommendations provided in the following sections. Several of these are targeted toward the USEPA. However, the workgroup emphasizes that, while the values and regulatory policies of international entities may vary widely, the fundamentals of regulatory science (toxicology, chemistry, and environmental fate) that form the basis for these recommendations are broadly applicable.

Short-term recommendations

Several improvements to current regulatory practice could be adopted within one year. The workgroup suggests the following:

1) Allow regulatory authorities the flexibility to use either appropriately sensitive toxicity tests (*e.g.*, WET) or numerical limits on metals, or both, when setting standards or making permitting decisions. Given the knowledge that site-specific factors can significantly reduce the toxicity of a metal, the reliance on only numerical limits to determine when a metal quality criterion is exceeded may not always be justified. Sufficient information is currently available to define the circumstances when toxicity-based controls can be used instead of, or in combination with, numerical limits.

2) Apply a tiered method using both chemical and biological tests to decide whether to regulate an effluent. The current U.S. process for considering site-specific information when deciding whether to regulate a metal relies on the WER, which can pose a significant financial burden on small dischargers. A tiered approach that starts with simple screening and then progresses only as needed to increasingly more comprehensive and costly information offers a better way to make these decisions in a sound and cost-effective manner. The workgroup recommends a tiered approach (Figure 3-1) that accomplishes the following:

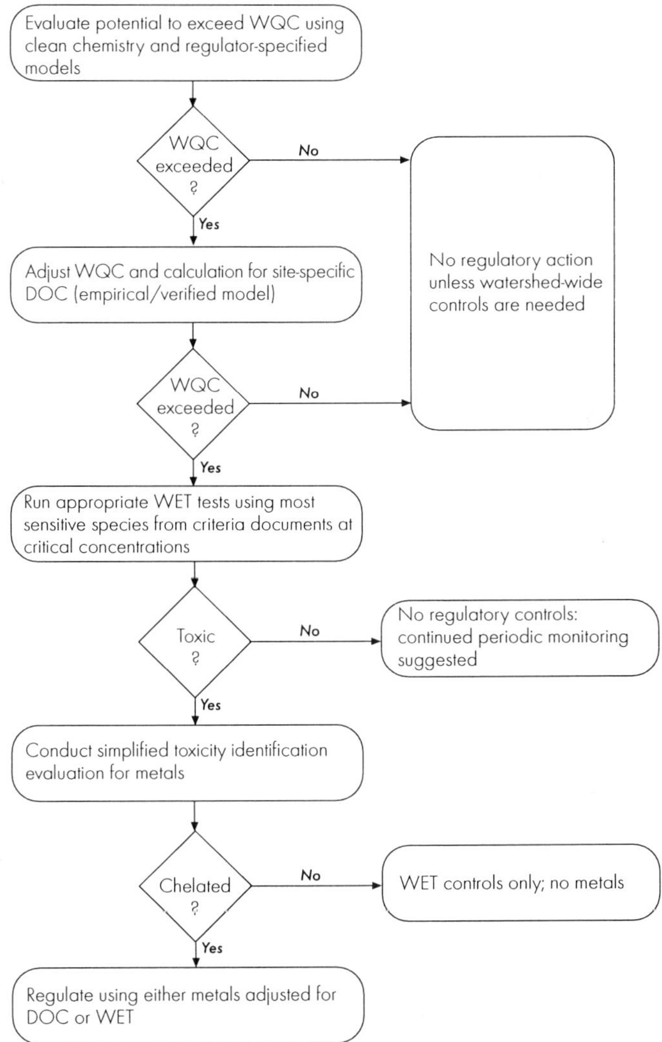

Figure 3-1   Recommended tiered procedure for regulatory decisions

- Starts with an evaluation of the potential to exceed a WQC concentration, based on data that are collected using clean chemistry and the permitting authority's dilution/loading models

- If necessary, continues with a site-specific adjustment to the WQC and model calculations based on an empirical DOC partitioning model or on a verified mechanistic model, if available

- If necessary, continues with laboratory toxicity tests of the effluent using organisms known to be sensitive at the criterion concentration, based on the data supporting the original criterion and run at conditions similar to "critical" (*e.g.*, dilution) conditions

- If necessary, ends with an abbreviated TIE using a metal chelator to confirm that metals are a source of toxicity

3) Evaluate existing data to determine the overall prevalence of actual environmental problems caused by metals for comparison with the prevalence of measurable problems caused by other factors, such as habitat destruction or nutrient enrichment. As with all societal resources, environmental agency resources are limited. Care should be taken to direct these resources toward the environmental problem areas where the risk is greatest. This evaluation also needs to consider the inherent uncertainties in collecting representative data and in making credible inferences. Finally, post-action monitoring and overall watershed assessments are needed to assess the cumulative effects of metal loadings.

4) Develop a database summarizing the experience of users of the WER approach in deriving site-specific WQC. Determine the relationships between WERs based on dissolved and total recoverable metal relative to water body type, effluent type, relative species sensitivity, and water quality factors. Determine the relationships among national WQC, WET tests, biocriteria, TIEs, and site-specific WQC derived using the WER. Identify the actual costs of these evaluation procedures.

5) Establish and maintain a high-quality database for metal concentrations in effluents and ambient waters. Much of the data in current monitoring databases on metals concentrations in effluents and ambient waters were not collected or analyzed using "clean" techniques, and the quality of these data is highly questionable. Regulatory and assessment decisions should not be made based on questionable data. It is recommended that a panel of experts be convened to derive protocols to assess the available data and that the current databases be corrected to contain only reliable and accurate information.

6) Clarify the requirement and intent of the USEPA's policy of independent application of standards. This policy has been widely interpreted as an inflexible requirement regarding the expression and measurement of standards attainment by each of three independent monitoring approaches (chemical-specific, WET, bioassessment). The scientific basis for extending this policy to criteria development and standards setting has been poorly understood; thus, the interpretation and application of the policy have been highly variable between federal and state regulatory authorities. The USEPA's interpretation should be clearly communicated to all regulatory agencies.

7) Emphasize the ability of permit regulators to incorporate seasonally dependent effluent limits in effluent discharge permits. Permit limits for some conventional pollutants should accommodate seasonal changes in environmental conditions, resulting in different permit limits for different seasons. This

flexibility can extend to toxic pollutants without compromising environmental protection and will result in permit limits that more closely consider the environmental conditions on which they are based.

Long-term recommendations

The following recommendations target research or additional information needs. The workgroup believes that the necessary research can be completed over the next three to five years.

1) Harmonize the chemical-specific, WET, and field biological assessment approaches. Current U.S. regulatory policy on Independent Applicability discourages the use of scientific principles for reconciling disparate results from these three methods. As a result, the policy discourages states from developing WET and bioassessment programs. It is recommended that research be initiated to develop the principles for including all information in regulatory decisions. This will involve developing a procedure for weighing all the evidence.

2) Develop a toxicity and chemical speciation model, validate it with multiple datasets, and use it to replace the current criteria development approach. This proposed model, or suite of similar models, will go to the heart of the toxicity problem by addressing the effects of metals on the sites where toxicity is expressed, while incorporating the influences of site-specific water chemistry. As a result, these models will make the new WQC directly applicable to individual waters without relying on the current approach of using site-specific studies. These models should be thoroughly tested for one metal, using multiple freshwater and marine organisms, both vertebrate and invertebrate, before being tested for all metals.

3) Quantify the actual cumulative safety factor in the regulatory decision process, and make it apparent to everyone. The current regulatory process includes safety factors that are both implicitly and explicitly introduced throughout the development of criteria, standards, TMDLs, and permits. These factors should be quantified and made apparent to decision-makers so that the true degree and sources of safety are known at the time of the decision. In addition, these factors should be documented and made part of the public record developed to support the decision.

Other recommendations

In addition to the short- and long-term recommendations in the preceding two sections, the workgroup identified other, lower priority recommendations:

1) Determine when ultraclean sampling and analytical methods for metals should be used in assessing fresh waters, marine waters, and effluents. The section titled "Lack of clean sampling and analytical techniques" indicates that clean analytical methods and techniques are essential for credible measurement of metal concentrations below the one-part-per-million level. To measure concen-

trations that are below the one-part-per-billion level, ultraclean methods may be necessary.

2) Reevaluate the NPDES rule that requires permit limits to be expressed in the total recoverable metal form, assuming that a sufficient number of models of dissolved metals support this change. This rule change should be considered only after sufficient field studies are conducted to test whether the dissolved fraction of metals in effluents is relatively independent of the particulate form. Some field studies suggest this, but there are an insufficient number of studies to support this conclusion.

3) Determine default values for translators that convert the total recoverable metal concentration to the dissolved concentration. Current methods rely considerably on site-specific information, which may not always be available or easy to obtain.

4) Determine whether meeting WQC in the water column is also protective of sediments. This is a question that dominates some discussions about the use of dissolved metal in establishing WQC, and it should be answered. This may require collection of new data, and thus the workgroup sees this as a long-term recommendation, although modeling could quickly shed some light on this issue.

5) Develop predictors of dissolution kinetics for metal dissolving from metal-based products. The workgroup sees a need to collect this information for hazard identification and labeling needs in the EU.

References

Adeloju SB, Bond AM. 1985. Influence of laboratory environment on the precision and accuracy of trace element analysis. *Anal Chemistry* 57:1728.

Allen HE, Hansen DJ. 1996. The importance of trace metal speciation to water quality criteria. *Water Environ Res* 68:42–54.

Ausley LW. 1996. Personal communication. North Carolina Division of Environmental Management, 4401 Reedy Creek Road, Raleigh NC 27607-6445.

Berman SS, Yeats PA. 1985. Sampling of seawater for trace metals. *CRC Reviews in Analytical Chemistry* 16.

Bloom NS. 1993. Ultra-clean sampling, storage, and analytical strategies for the accurate determination of trace metals in natural waters. Presented at the 16th annual EPA Conference on the Analysis of Pollutants in the Environment; 1993 May 5; Norfolk VA.

Cappel CR. 1996. Personal communication. Eastman Kodak Company, 69 Hermitage Road, Rochester NY 14617.

[CCME] Canadian Council of Ministers of the Environment. 1991. A protocol for the derivation of water quality guidelines for the protection of aquatic life. In: CCREM 1987, Appendix IX. Winnipeg, Manitoba.

[CCME] Canadian Council of Ministers of the Environment. 1993. Protocols for deriving water quality guidelines for the protection of a agricultural uses. In: CCREM 1987, Appendix XV. Winnipeg, Manitoba.

[CCME] Canadian Council of Ministers of the Environment. 1996a. A protocol for the derivation of environmental and human health soil quality guidelines. CCME Subcommittee on Environmental Quality Criteria for Contaminated Sites. Winnipeg, Manitoba.

[CCME] Canadian Council of Ministers of the Environment. 1996b. (Draft). A protocol for the derivation of Canadian tissue residue guidelines for the protection of wildlife that consume aquatic biota. Prepared by the CCME Task Group on Water Quality Guidelines. Winnipeg, Manitoba.

[CCME] Canadian Council of Ministers of the Environment. 1996c. Protocol for the derivation of Canadian sediment quality guidelines for the protection of aquatic life. Prepared by the CCME Task Group on Water Quality Guidelines. Winnipeg, Manitoba.

[CCREM] Canadian Council of Resource and Environment Ministries. 1987. Canadian Water Quality Guidelines. Prepared by the Task Force on Water Quality Guidelines of the Canadian Council of Resource and Environment Ministers.

Council Directive 76/464/EEC. 1976. *Off J Eur Communities* L129/23.

[CWA] Clean Water Act. 33 U.S.C. §1251 *et seq.* (June 30, 1948). Also titled Federal Water Pollution Control Act.

Davies T. 1991. Transmittal memorandum of final policy on biological assessment and criteria. Memorandum to Water Management Directors, USEPA Regions I-X. Washington DC: Office of Water, USEPA.

Gaudet CL, Keenleyside KA, Kent RA, Smith SL, Wong MP. 1995. How should numerical criteria be used? The Canadian approach. *Human Ecol Risk Assess* 1:19–28.

Grothe DR, Dickson KL, Reed-Judkins DK, editors. 1996. Whole effluent toxicity testing: an evaluation of methods and prediction of receiving system impacts. SETAC Pellston Workshop on Whole Effluent Toxicity; 1995 Sep 16–25; Pellston MI. Pensacola FL: SETAC Pr. 350 p.

Hudson-Edwards KA, Macklin MG, Curtis CD, Vaughn DJ. 1996. Processes of formation and distribution of Pb-, Zn-, Cd-, and Cu-bearing minerals in the Tyne Basin, northeast England: implications for metal-contaminated river systems. *Environ Sci Technol* 30:72–80.

Perwak J, Bysshe S, Goyer M, Nelken L, Scow K, Walker P, Wallace D, Delos C. 1980. An exposure and risk assessment for copper. Washington DC: USEPA Office of Water Regulations and Standards. National Technical Information Service (NTIS) No. PB8 5 211985/AS.

Prothro MG. 1993. Office of water policy and technical guidance on interpretation and implementation of aquatic metals criteria. Memorandum from Acting Assistant Administrator for Water. Washington DC: USEPA Office of Water. 7 p. Attachments 41 p.

Seventh Amendment of 1992. Amendment of Council Directive 67/548/EEC. *Off J Eur Communities* L154.

Spliethoff HM, Hemond HF. 1996. History of toxic metal discharge to surface waters of the Aberjona watershed. *Environ Sci Technol* 30:121–128.

[TSCA] Toxic Substances Control Act. 15 U.S.C. §2601–2692. (October 11, 1976).

[USEPA] United States Environmental Protection Agency. 1983a. Overall margin of protection and margin of safety related to development of wasteload allocations. Washington DC: USEPA Office of Water, Regulations and Science. (draft).

[USEPA] United States Environmental Protection Agency. 1983b. Water Quality Standards Handbook. Washington DC: USEPA Office of Water, Regulations and Standards (WH-585).

[USEPA] United States Environmental Protection Agency. 1984a. National pollutant discharge elimination system permit regulations. *Federal Register* 49:38028–38029.

[USEPA] United States Environmental Protection Agency. 1984b. Guidelines for establishing test procedures for the analysis of pollutants. *Federal Register* 49:43431.

[USEPA] United States Environmental Protection Agency. 1986. Quality criteria for water. Washington DC: USEPA Office of Water Regulations and Standards. EPA 440/5-85-001.

[USEPA] United States Environmental Protection Agency. 1990. Biological criteria: national program guidance for surface waters. Washington DC: USEPA Office of Water. EPA/440/5-90/004.

[USEPA] United States Environmental Protection Agency. 1993. Water quality criteria: aquatic life criteria for metals. *Federal Register* 58(108):32131–32133.

[USEPA] United States Environmental Protection Agency. 1994a. Contaminated sediments management strategy. Washington DC: USEPA Office of Water. (proposed).

[USEPA] United States Environmental Protection Agency. 1994b. Interim guidance on determination and use of water effects ratio for metals. Washington DC: USEPA Office of Science and Technology. EPA/823/B-94/001.

[USEPA] United States Environmental Protection Agency. 1995a. Guidance on establishing trace metal clean rooms in existing facilities. Washington DC: USEPA Office of Water. EPA/821/R-95/034.

[USEPA] United States Environmental Protection Agency. 1995b. Sampling ambient waters for trace metals at EPA water quality criteria levels. Washington DC: USEPA Office of Water. EPA/821/B-95/001.

[USEPA] United States Environmental Protection Agency. 1996. The metals translator: guidance for calculating a total recoverable permit limit from a dissolved criterion. Kinerson RS, Mattice JS, Stine JF, editors. Washington DC: USEPA Office of Water. EPA 823-B-96-007. 60 p.

Chapter 4

Environmental toxicology of metals

Christopher M. Wood, chair
William J. Adams, Gerald T. Ankley, Donald R. DiBona,
Samuel N. Luoma , Richard C. Playle, William A. Stubblefield,
Harold L. Bergman, Russell J. Erickson, Jack S. Mattice, Christian E. Schlekat

Water quality criteria to protect aquatic life do little to address the likely effects of water quality on the toxicity of metals in the environment. Only in the case of a few metals have any bioavailability-altering environmental factors (*e.g.*, water hardness, as well as water quality conditions that may be correlated with hardness such as alkalinity) been considered in developing criteria. However, a variety of other factors, *e.g.*, pH, chloride, and DOC, can also control bioavailability of metals. In an attempt to resolve or at least reduce these concerns, criteria approaches have been modified, *e.g.*, the use of "dissolved" rather than "total recoverable" measurements. The addition of the ability to use dissolved metal instead of total recoverable metals values for permits, when appropriate, is a step in the right direction. However, criteria approaches are not yet based on a good theoretical understanding of the reasons why metal toxicity varies with water quality. Indeed, the mechanisms of toxicity of some metals and the ability to predict metal speciation and binding of metals to DOC have only recently been demonstrated. As these mechanistic approaches for handling metal speciation, binding, and toxicity estimation develop, they should be incorporated into the guidelines for WQC and WQS.

Water effect ratios: the interim fix

The original guidance for deriving national WQC (USEPA 1983) describes the advantages of developing site-specific criteria to reflect species of local concern and the effects of local conditions on toxicity. Guidelines for deriving site-specific criteria were developed and subsequently updated by the USEPA (1984, 1992, 1994), providing a series of protocols for modifying national WQC to reflect local environmental conditions. These guidelines are quoted below.

> "Site-specific criterion derivation may be justified because species at the site may be more or less sensitive than those in the national criterion document," or because "...differences in physical and chemical characteristics of water have been demonstrated to ameliorate or enhance the biological availability and/or toxicity of chemicals in freshwater and saltwater environments."

Site-specific criteria are intended for use by state regulatory agencies responsible for developing and enforcing WQS, mixing zone standards, or toxicity-based effluent standards. When implementing these standards, USEPA guidance (1984) recommends that the states take into account such factors as these:

- Intended use of a given site
- Social, legal, and economic ramifications of the severity of the standards

- Environmental and analytical chemistry of the material (*i.e.*, how the chemical behaves in ambient waters, whether it is bioavailable, and whether certain analytical methods accurately predict the toxicologically effective chemical concentration)
- Strength of any extrapolations relating laboratory testing data to site situations and population effects
- Relative intrinsic/economic value and sensitivities of the species for which toxicity data are available versus those of species resident in the water body that is to be protected

The national criteria for metals generally are based on laboratory data reported as total, total recoverable, or acid extractable metals concentrations; the criteria, therefore, are expressed in terms of "total recoverable" metals concentrations. Although this standardization simplifies the application of metals criteria to various water bodies, the speciation of a number of metals, which may vary tremendously depending on the chemical and physical make-up of a given natural water, strongly alters the toxicological properties and biological availability of those metals. According to the USEPA (1984), national WQC may be too conservative when applied to total metals concentrations in waters where the majority of the metal is not in a dissolved form (*i.e.*, contained within or bound to suspended particles) or is strongly complexed (*i.e.*, to organic or other ligands). Implementing procedures for deriving site-specific criteria is encouraged "for those metals whose biological availability and/or toxicity is significantly affected by variation in physical and/or chemical characteristics of water" (USEPA 1984).

It is clear from the existing data and is recognized in the USEPA's policy documents that national WQC for metals can be either over- or underprotective under certain environmental conditions. Measurement of the site-specific WER[1] allows for the determination of an appropriate adjustment factor that renders the national criteria more applicable to the water body under consideration. By determining the ratio between metals toxicity in the actual site water and in laboratory water, the national criterion value can be adjusted to more clearly reflect bioavailability at the site.

Beginning with the original Guidelines for Deriving Numerical National Water Quality Criteria for the Protection of Aquatic Organisms and their Uses (Stephan *et al.* 1985), and followed by the Guidelines for Deriving Numerical Aquatic Site-Specific Water Quality Criteria by Modifying National Criteria (Carlson *et al.* 1984), the Interim Guidance on Interpretation and Implementation of Aquatic Life Criteria for Metals (USEPA 1992), the Interim Guidance on the Determination and Use of Water-Effect Ratios for Metals (USEPA 1994), and the Interim Final Rule for the Establishment of Numeric Criteria for Priority Toxic Pollutants (USEPA 1995), the USEPA has presented a clear policy recognizing the technical value of site-specific criteria modification and has promoted it under specified circumstances.

---

[1] Water-effect ratio is calculated from pairs of side-by-side toxicity tests as the quotient of the two toxicity endpoints (*e.g.*, WER = the median lethal concentration [LC50] in site water divided by the LC50 in laboratory water). Water-effect ratios are used to directly modify national criteria for site-specific application.

Additionally, USEPA interim guidance allows for the dissolved form of the metal to be used as a basis for expressing standards and setting permit limits (Prothro 1993); this interim guidance is believed to provide a more accurate way of assessing the potential for site-specific toxicity. Recently, guidance has become available on translating between total and dissolved forms (Elseroad *et al.* 1994; USEPA 1996).

The WER approach has several tangible benefits over previous approaches to criteria. While this method has the benefit of incorporating an empirically observed value into the criteria approach, it will remain ambiguous until a clear understanding of the mechanisms affecting metals bioavailability is obtained. This shortcoming, although troubling, should not discourage the use of the technique; rather, it should encourage the development of alternative mechanism-based approaches such as have been proposed in this workshop. Site-specific, WER-modified dissolved metals criteria provide an improvement over blanket application of national WQC. Until such time as mechanistic models are available for regulatory implementation, steps should be taken to further refine the current WER determination procedures (USEPA 1994), and steps should be taken to ensure that USEPA regional staff and state permit writers fully understand the basis, intent, and use of the procedure and are able to apply it in a uniform manner.

Generation of metals criteria using a receptor loading model: a new approach

In the mid-1980s, when many of the USEPA's AWQC for metals were developed using data generated in the preceding two to three decades, there was limited understanding of the mechanisms of metal toxicity for both acute and chronic effects. It was generally believed that the acute toxicity of metals affected fish, for example, by damaging gill function with explanations such as "suffocation," "generalized edema," and "mucification" being commonly presented (Skidmore and Tovell 1972; Mallat 1985). Chronic toxicity was attributed to the debilitating internal effects of bioaccumulation. From the standpoint of acute effects, the inherent assumption was that different metals had similar mechanisms of toxicity but different potencies. Since that time, mechanisms of chronic toxicity have undergone further study, but the relationship between bioaccumulation and toxicity is still unclear and mechanisms of chronic toxicity remain poorly understood (McDonald and Wood 1993).

Understanding mechanisms of acute toxicity
Mechanisms of acute metal toxicity. In the past decade, an abundance of mechanistic studies has greatly expanded understanding of how metals damage aquatic organisms at acute concentrations. In freshwater organisms in particular, at exposure concentrations producing acute effects, these studies have confirmed that the gills are the principal site of toxicity (Evans 1987; Wood 1992). Gills normally serve two critical functions in freshwater fish: gas exchange (O_2, CO_2, NH_3) and active ion uptake (principally Na^+, Cl^-, Ca^{2+}) to counter ion losses down their electrochemical gradients from the fish to the water. These studies have shown the following:

- Metals such as copper, zinc, cadmium, silver, and aluminum generally act as "surface active" toxicants, exerting their damaging effects by binding to anionic sites on or in the gills. These sites play critical roles in gill transport functions. The physiological consequences of these deleterious surface effects are far more serious than any systemic effects resulting from actual entry and accumulation of metal within the organism.

- Free metal cations generally appear to be the toxic metal species (but see discussions of dietary exposure/toxicity in the section titled "Understanding multiple exposure routes").

- Different metals have different mechanisms of toxicity that relate to interference with different specific gill functions as a result of metal binding.

Current understanding of ion and gas exchange processes in the freshwater fish gill is illustrated in Figure 4-1.

The major actions of most metals in the low μg/L range may be understood by reference to this model (see Table 4-1).

It is therefore clear, at least for freshwater fish, that the metals studied in mechanistic detail to date exert their toxicity largely by blocking one of two essential gill transport functions: Na^+ uptake by copper, silver, and aluminum or Ca^{2+} uptake by cadmium and zinc. Potentially lethal decreases in plasma Na^+ concentrations (*hyponatremia*) or Ca^{2+} concentrations (*hypocalcemia*) result. Aluminum is unique in causing respiratory toxicity at environmentally relevant concentrations due to gill swelling, in addition to its effect in inhibiting Na^+ uptake.

The knowledge that metals exert their toxic effects by binding to specific transport sites on the receptor surface of the organism opens up the possibility of a fundamentally new approach to water quality regulation. This new approach incorporates the key receptor sites on or in the organism as the response surface for toxicity. Ultimately, this approach holds the promise of application not only to acute toxicity in fish but also to acute toxicity in invertebrates and to chronic toxicity.

Metal binding in fish gills. In the last three years, three discoveries indicate how the new information on acute toxic mechanisms might be used to construct a more scientifically defensible approach to regulating metals in the environment:

1) The different gill binding sites have been quantified and differentiated in terms of affinity ("apparent binding constants" or "conditional stability constants" [Playle *et al.* 1993b; MacRae *et al.* 1997]).

2) The available data suggest that metals bind to these sites in a predictable fashion that is amenable to geochemical modeling. This modeling employs currently available computer programs (*e.g.*, MINEQL$^+$ [Schecher and McAvoy 1992], MINTEQA2 [Allison *et al.* 1991]) and takes into account competing reactions between the organism's binding sites and all other anionic ligands for the metal in the water column. In addition, the models consider competition between all other cations in the water column and the metal for the binding

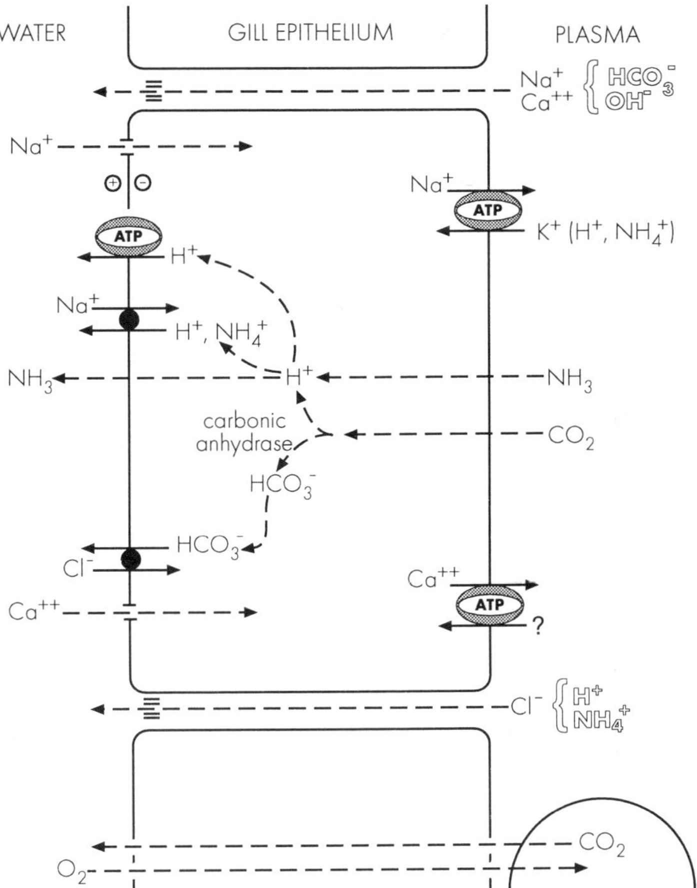

Figure 4-1   A functional model of ionic transport mechanisms in the gills of freshwater fish. Source: Wood 1992. Reprinted from *Aquatic Toxicology* 22: 239–264, 1992, with kind permission of Elsevier Science –NL, Sara Burgerhartstraat 25, 1055 KV Amsterdam, The Netherlands.

sites on the organism (Playle *et al.* 1993a; Janes and Playle 1995; MacRae *et al.* 1997).

3) Limited available data suggest that relative saturation of the specific binding sites by the metal (*i.e.*, gill metal burden or residue) can be used as an accurate predictor of eventual mortality or survival (MacRae *et al.* 1997).

It is now possible to consider development of a regulatory strategy based on the response surface of the organism as the meaningful biological receptor (Figure 4-2). The degree of saturation of that surface with metal (*i.e.*, the tissue residue or load) is directly linked to a meaningful biological endpoint (*i.e.*, death). The linkage occurs by a known physiological mechanism, in contrast to previous approaches, which considered neither the mechanism of action nor the true bioavailability of the metal.

Table 4-1   Physiological mechanisms of metal toxicity in freshwater fish at levels representative of the acute criteria

Metal	Toxic Effect	Reference
Cu	Blockade of Na^+ uptake	Laurén & McDonald (1986)
		Wilson & Taylor (1993)
		Pilgaard *et al.* (1994)
Ag	Blockade of Na^+ uptake	Wood *et al.* (1996)
Cd	Blockade of Ca^{2+} uptake	Verbost *et al.* (1987)
Zn	Blockade of Ca^{2+} uptake	Spry & Wood (1985)
		Hogstrand *et al.* (1994)
Al	Blockade of Na^+ uptake	Staurnes *et al.* (1984)
		Playle *et al.* (1989)

The binding constants of the response surface relative to the binding constants of the various anions and cations in the water column can be used to predict the answer to the fundamental question: "Is the metal going to be bioavailable in this particular effluent or in a receiving water of known composition?" If necessary, the predicted answer can be independently validated by exposing organisms under field conditions and measuring the gill metal burden. Because the organism itself serves as the quantified receptor, the approach also holds the promise of greater sensitivity and accuracy under real world conditions. The concept may be applicable to all routes of exposure (*e.g.*, dietary) so long as the tissue load is measured on the appropriate receptor surface. While this approach appears theoretically sound, a great deal more work will be needed to validate it under a wide range of environmental and toxicological conditions. The modeling approach is outlined in greater detail in the section that follows.

Modeling metal-gill interactions

It is assumed that metal toxicity to aquatic organisms occurs as a result of metal-ligand interactions. That is, a metal must first bind to the surface of a target organ before toxicity occurs at that site or at other sites after internal transport of the metal. If a metal does not bind to the site of uptake, it will not be toxic. Gills of fish are the first organs affected by many waterborne metals and will be used to illustrate the use of metal-gill binding to model metal toxicity in aquatic organisms.

Individual metals bind to specific receptor sites on the gills and cause specific physiological damage that may lead to mortality. Toxicities of metals may be reduced in water with higher $Ca2^+$ and DOC concentrations, and H^+ ions usually protect against these metals (but are toxic themselves). A good conceptual model for these protective effects is the "gill surface interaction model" first presented by Pagenkopf (1983) and then by Morel (1983) and Morel and Hering (1993).

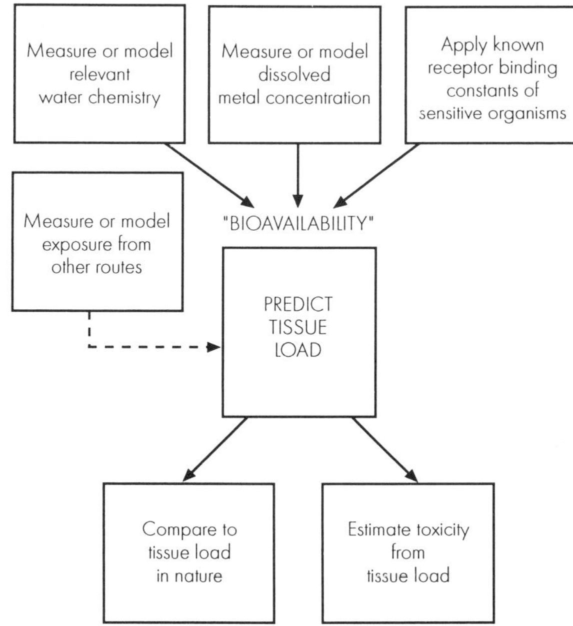

Figure 4-2   Use of receptor loading model to predict metal bioavailability

The gill surface interaction model considers acutely toxic metals that alter gill functions. In this model, the toxic species of a metal is considered to be the "free metal" or cationic form (*e.g.*, Cu^{2+}, Cd^{2+}). Chemical reaction rates are fast in the model, and the gills have a finite number of negatively charged binding sites to which metal cations bind. Competition for the gill binding sites occurs between metals, H^+ ions, and physiologically important cations such as Na^+ and Ca^{2+}. Finally, the number of free metal ions is reduced through complexation with aquatic ligands such as DOC, HCO_3^-, Cl^-, *etc.*, or with synthetic ligands as are sometimes used in laboratory experiments (*e.g.*, ethylenediamine tetraacetic acid [EDTA]).

A metal kept off the gills through competition (*e.g.*, by Ca^{2+} or H^+) or complexation (*e.g.*, by DOC) will not be toxic to a fish. If the binding strength of a metal to the gill can be calculated and the number of binding sites determined, then competition and complexation reactions can be treated mathematically. A predictive tool can then be developed based on water chemistry as well as on dissolved metal concentration.

The basic approach to developing this model is to measure metal deposition on fish gills after exposure of the fish to the metal in a synthetic softwater system. The water is then modified by adding cations such as $Ca2^+$, Na^+, and H^+ (competition experiments) or various ligands such as EDTA or natural DOC (complexation experiments). By running these types of experiments using synthetic ligands of known metal binding strength, a metal-gill conditional stability constant can be calculated (the "ligand exchange" method). This mathematical approach has been applied to algae (*e.g.*, Sunda and Guillard 1976; Xue and Sigg 1993) and was used as early as 1981 with copepods (Borgmann 1981) and in a series of studies (Daly *et al.* 1990a, 1990b, 1990c, 1992) using a freshwater shrimp.

Modeling results for copper binding at fathead minnow gills (Playle *et al.* 1993a, 1993b) will be used to illustrate this approach. Fathead minnows were exposed to copper for 3 hours in synthetic soft water; then their gills were analyzed for copper. If copper was kept off the gills in the presence of a synthetic ligand of known binding strength, then the binding strength of copper to the gills could be calculated. For example, copper binds to EDTA with a conditional binding constant (K) of $10^{18.8}$, $\log K = 18.8$. If EDTA keeps copper off the gills by complexing the copper, then the $\log K$ value for copper binding to the gills is less than 18.8. From experiments such as this, Playle *et al.* (1993a, 1993b) were able to calculate a Cu-gill binding constant of 7.4, with 2 nmol binding sites per fish. From this value, they were able to run competition experiments with Ca^{2+} and H^+ and to calculate Ca^{2+} and H^+ binding constants at the copper binding sites ($\log K = 5.4$ and 3.4, respectively; Figure 4-3). The binding constant of copper to DOC was calculated as $\log K = 9.1$, with about 50 nmol binding sites per mg $C \cdot L^{-1}$ DOC.

Using these values inserted into the MINEQL$^+$ aquatic chemistry program, Playle *et al.* (1993a) were able to predict copper binding to gills of minnows exposed to copper in a series of lake and river waters. Waters with high Ca^{2+} and DOC concentrations kept Cu^{2+} off the gills better than did waters containing less Ca^{2+} and DOC. In similar experiments using brook and rainbow trout, MacRae *et al.* (1997) found $\log K$ Cu-gill binding constants of 7.2 and 7.5, respectively (*cf.* 7.4, Figure 4-3), which indicates that the approach produces similar $\log K$ values.

This approach directly considers the gill as the target organ of a metal, and it creates an interface between powerful geochemical speciation models and the organism that can be affected by the metal. It is a step beyond models such as that which Welsh *et al.* (1993) developed for larval fathead minnows. That correlational approach considered water pH and DOC concentrations in an empirical model to predict 96-h LC50 for copper values. Inserting equilibrium binding constants for metal-gill binding into an aquatic chemistry program allows much more complex water chemistry to be taken into account when predicting toxic effects of metals. For example, complexation of copper by HCO_3^- is explicitly considered by the MINEQL$^+$ computer program.

Modeling metal-gill interactions has been done for copper and cadmium (Playle *et al.* 1993a, 1993b), silver (Janes and Playle 1995), and lead, cobalt, and mercury (Richards *et al.* unpublished). Generally, the extent to which metals bind to the gills is in accordance with their strength of binding to synthetic ligands such as EDTA. Exceptions to the expected binding order are interesting in that they might reflect physiological mechanisms at the gills. For example, the stronger than expected binding of cadmium to the gills (Playle *et al.* 1993a) probably reflects binding of cadmium at high affinity Ca^{+2} channels (Verbost *et al.* 1989). Thus, good knowledge of physiological effects of metals is necessary to properly interpret model results.

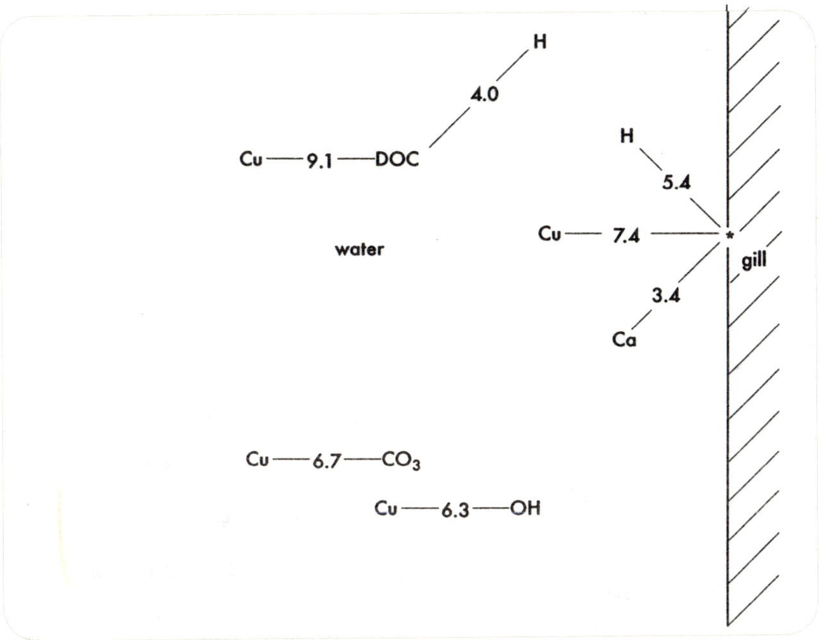

Figure 4-3   Illustration of the most important parameters in the Cu-gill model (Playle *et al.* 1993b). Complexation of Cu in the water and competition with cations at a Cu binding site (asterisk) can reduce Cu binding at the gills.

Although this approach to modeling the effects of metals in the aquatic environment promises to be an extremely valuable tool, it requires further development. Future research needs to include more studies done in a range of different water chemistries with differing pH, calcium concentrations (hardness), bicarbonate concentrations (alkalinity), DOC concentrations, and salinities. Depending on the metal and on the buffer capacity of the water in question, pH changes in water near the gill surface may also need to be considered: poorly buffered acidic water is rendered more basic, and poorly buffered basic water is made more acidic as it passes over the gills (Playle and Wood 1989; Lin and Randall 1990; Playle, Appendix). Sensitivity analysis using the model will help determine which water chemistry parameters are most important in a particular water body.

In addition, more work needs to be done to correlate gill metal burden with metal toxicity. Limited data available on copper binding to trout gills (MacRae *et al.* 1997) show that metal deposition is directly related to toxicity, and mixed copper and cadmium toxicity was correlated with gill cadmium concentrations in fathead minnows (Playle *et al.* 1993a). However, more data are needed and field validation is ultimately required.

The applicability of the model to metal mixtures needs to be assessed. A first step is determining whether particular metals bind at the same sites or different sites. For example, copper affects Na+ uptake and cadmium affect Ca^{2+} uptake in freshwater fish gills (Laurén and McDonald 1986, 1987; Verbost *et al.* 1989), so they are assigned different binding sites in the metal-gill model of Playle *et al.* (1993a). The degree to which effects are additive, whether at the same site or different sites, must then be determined. For chronic exposures, additional considerations would include detoxification mechanisms, such as how the induction and protective effects of metallothionein are affected by exposure to multiple metals.

Modeling metal binding to gills of fish directly considers the influence of water chemistry on metal binding at the target organ of acute metal exposures in fish. This approach explicitly links water chemistry, metal concentration, and the organism.

Extending the proposed approach to invertebrates

For several of the marine and freshwater metal criteria, invertebrates (in particular, juvenile and larval life stages) have been among the most sensitive organisms tested. Unfortunately, an understanding of mechanisms of acute toxicity and, hence, sites of action of metals generally is lacking for invertebrates. The proposal that metal criteria for fish could be linked to tissue-specific residue concentrations (*i.e.*, gills for acute lethality, "receptor-loading model") is, from a mechanistic standpoint, predicated upon the fact that the tissue selected for analysis is the site of action for expression of toxicity (*i.e.*, for the fish gill, ionoregulatory and respiratory stress). Given this, it is clear that serious consideration should be given to how this criteria approach might be applied to animals such as invertebrates, where there is a lack of understanding of metal toxicity mechanisms.

One approach would be to develop a toxicology program focused upon defining and validating toxic modes of action of metals for those invertebrate species/genera identified as most sensitive in WQC documents for specific metals. This would enable accurate definition, for example, of toxic sites of action for metals and would optimize successful development of tissue residue-toxicity relationships. Although technically rigorous, this approach could prove very challenging given the current lack of understanding of basic physiology (*e.g.*, osmoregulation, respiration, waste excretion) in many invertebrates, not to mention the logistical challenges associated with factors such as small sample mass.

In the case of invertebrates, the use of tissue residue-toxicity relationships based on whole body residues might be more practical and adequately predictive if metal concentrations in the whole animal co-vary with metal concentrations at the site of action. That is, it may not be necessary to completely understand mechanisms of metal toxicity in invertebrates. If this is true, it should be possible to use purely empirical relationships between toxicity and total body burdens of metals to derive metals criteria for invertebrates.

Building upon this construct, the following strategy is proposed. First, collect and evaluate existing metal tissue residue-toxicity data for metals in invertebrates. To this end, researchers at the Duluth USEPA laboratory have compiled an exhaustive database (*ca.* 500 references) of tissue residue-toxicity relationships for a variety of chemicals, including pesticides, metals, and industrial narcotics in fish and aquatic invertebrates. Work has also been done at the University of Wyoming relating metal partitioning to the cladoceran *Daphnia magna*, relative to water quality characteristics, in a manner analogous to that used for the fish gill model (Boese and Bergman unpublished). After existing data are evaluated for potential utility and hypothesis formulation, actual testing could be done with sensitive invertebrates (identified from WQC documents) to supplement existing data concerning empirical relationships between total body metal residues and acute toxicity. As for the fish gill analyses, these tests should be done under variable water quality conditions (calcium, sodium, pH, DOC, *etc.*) designed to evaluate the relative utility of predicted/observed metal partitioning in the organism and toxicity. Depending on the results from compilation of existing data, as well as on evaluation of the basic metal residue-toxicity experiments, it may be deemed unnecessary to pursue a more mechanistic understanding of invertebrate sensitivity.

A number of potential uncertainties arise in using whole body metal residues as opposed to target tissue metal residues to predict effects in invertebrates. One of the foremost issues here would be that of equilibrium. For example, a very high metal exposure concentration acting at the respiratory/ionoregulatory surface might kill the organism quickly at total body burdens that might seem small, relative to those that occur in whole organisms exposed to slightly smaller metal concentrations for a longer time. Even if the cause of toxicity were similar in the two circumstances (acute respiratory-ionoregulatory distress), because of variable kinetic considerations the organisms exposed to lower concentrations may exhibit higher total body burdens, even though final toxic residues at the site of action were similar. Another potential problem with using total body residue-toxicity relationships for invertebrates is related to metal sequestration in the organism. In addition to metal detoxification through mechanisms such as metallothionein production, several invertebrates reduce the biological availability of accumulated metals through deposition of metals in their exoskeleton or carapace, yet contributions from these parts of the body would be measured in the total body burden of metal. Moreover, the extent of uncertainty arising from this deposition would depend upon the molting stage of the organism. A final complication that might arise in relating whole animal metal residues to effects (at least in the case of copper) is that some invertebrates (*e.g.*, crayfish) possess copper-based respiratory pigments. If any or all of these factors contribute significant uncertainty to tissue residue-toxicity relationships in invertebrates, it would be necessary to pursue mechanistic toxicology studies with select metals and biota.

Virtually all of the preceding considerations apply as well to chronic toxicity. For example, in the absence of a complete understanding of chronic toxicity mechanisms, derivation of robust tissue residue-effects relationships would be required. Also, interferences such as the induction of metallothionein, contributions to total metal body burden from the exoskeleton, *etc.*, would need to be considered. The one uncertainty that might be minimized relative to acute toxicity would be that of assuming an equilibrium between metal residues in target organs versus whole body residues.

Chronic toxicity

Chronic toxicity is presently defined by long-term laboratory exposures designed to assess endpoints important to the protection of populations and communities. Older literature exists that reports toxic concentrations for different life stages of bioassay species and has been used to calculate ratios to relate acute toxicity and chronic toxicity. Unfortunately, it appears that lower priority is given to studies of chronic toxicity at present than in the past. However, links between tissue residues and chronic toxicity are not well understood, specific mechanisms of chronic toxicity are not known, and the available data do not fully consider modes of chronic toxicity that might result in disappearance of populations and impairment of individual organisms in nature.

Links between tissue residues and chronic toxicity

Existing data are insufficient to understand linkages between chronic toxicity and tissue residues, either for gill metal loads or for more complex exposures (*e.g.*, dietary and other exposure pathways). In lieu of understanding such linkages, the acute-to-chronic ratios or other appropriate safety factors provide a reasonable interim approach for protecting ecosystems from chronic toxicity. Establishing linkages between tissue residues and mechanisms of toxicity (or surrogates of the responsible mechanisms) should be a high priority in the development of the tissue-residue, mechanism-based approach to regulation.

Specific mechanisms of chronic toxicity

Although not known, many benefits could accrue from better understanding of chronic toxicity mechanisms. Especially if a residue-based model ("receptor-loading model"), such as that discussed for the fish gill, is used to regulate metals, then the chronic manifestation of toxicity associated with the immediate mechanisms of toxicity must be better understood. At present, these relationships are not known.

Because the fish gill approach is a mechanism-based approach to regulation, the least ambiguous extrapolations from acute to chronic toxicity will involve extending our knowledge of mechanisms as a first step. For example, if the mechanisms of chronic toxicity are the same as those of acute toxicity, then predictions of one as a constant proportion of the other are warranted (*e.g.*, the present acute-to-chronic ratio). However, the mechanisms of the two may differ completely, or chronic toxicity may take the form of general stress reactions (loss of condition, growth impairment, reduced fecundity) that result when the mechanisms immediately affected by metals are disrupted for a long period of time. In these cases, the relationship between acute and chronic toxicity would be more complex, and the logic for extrapolating from acute to chronic effects would be weakened.

A second benefit of mechanistic knowledge might be a better understanding of why species differ in susceptibility to chronic exposure and, therefore, which species are likely to be most sensitive. For example, many cellular processes are conservative among species; therefore, it seems likely that commonalities will occur in acute responses to toxicity (*e.g.*, disruption of Na^+ or Ca^{+2} transport physiology by Cu^{2+} and Cd^{2+}, respectively, may be widespread among species). However, chronic toxicity also incorporates the capacity of organisms to adapt to cellular-level changes, and this could vary widely among species. Measuring the range of adaptive capacity could thus provide a basis for better understanding of interspecies differences in chronic metal sensitivity.

Protection of levels of biological organization relevant to ecosystems (*i.e.*, populations and communities) requires an understanding of the relationships between acute and chronic toxicity. Understanding mechanisms is the best approach to reducing ambiguities in such relationships. In addition to laboratory studies that link tissue loads with immediate stress responses, the following should also be considered:

a) Controlled study of links between tissue residues and general chronic stress responses

b) Controlled study of how long-term disruption of processes that are directly affected by metals (*e.g.*, disruption of Ca^{2+} and Na^+ transport) become manifested in chronic toxicity

c) Field study of the nature of relationships between tissue residues and signs (*e.g.*, biomarkers) in organisms, populations, and communities consistent with metal toxicity

Types of chronic toxicity that might influence populations and communities in nature

The challenge of acquiring better knowledge about chronic effects on populations and communities is substantial, as is the challenge of improving approaches to field validation. Better understanding of the mechanisms of chronic toxicity at the individual level might enhance prediction of life-cycle disruptions in populations. Better understanding, from mechanisms, of differences in vulnerability among species might enhance understanding of community changes (especially those difficult changes that have already occurred). More comprehensive toxicity tests are needed for important bioassay species that include long duration exposures, full life-cycle exposures, or multiple-generation exposures. Long-term, multidisciplinary study of contaminated ecosystems might aid in understanding the influence of metals compared to other disruptions or natural processes. As regulations are implemented or environmental cleanups are in progress, carefully designed and interpreted monitoring data might aid our understanding, not only of individual ecosystems but also of ecosystem recovery processes (and thus, of ecosystem disruption) in general.

All such studies should recognize established principles of toxicological action at multiple levels of organization (*i.e.*, the significant effect of a disruption occurs at the next level of organization up from an observation; the mechanisms should be studied at the next level down the scale of organization). They should also recognize the different time scales of toxicity. For example, for an animal with a generation time of one year, adaptation and/or disruption of hormones may take place on time scales of minutes; membrane process disruptions are manifested in hours; trophic effects may begin in hours, but take days to manifest themselves; growth or reproductive disruption may take months; disappearance or adaptation of the population will take several generations; effects on the community if the species is lost (and is a critical or keystone species) could take a decade or more to fully manifest themselves. These are complicated biological and ecological phenomena that are inherently difficult to understand, but achieving such understanding is critical to efficient and effective protection of ecosystems.

Understanding multiple exposure routes

Neither the water nor the sediment bioassay approaches currently used to determine WQC or SQC fully consider the effects of metal uptake via the digestive tract, as might occur when animals ingest metal-contaminated food (*e.g.*, fish, macroinvertebrates) or particles (*e.g.*, filter feeders, surface deposit feeders). If bioavailability via uptake from ingestion is additive with uptake from water, current regulations could underestimate metal effects, especially in circumstances where food is a predominant pathway. For regulatory purposes, it would be advantageous to develop a model that

1) predicts tissue metal concentration from multiple exposure routes, and
2) determines threshold tissue (*e.g.*, gill, gut, liver) concentrations which, if exceeded, result in toxicity.

Direct uptake of metals from food or particulate sources: an important exposure pathway for many organisms in many circumstances

Many early experiments with fish and invertebrates were ambiguous with regard to metal availability from food. The development of new methodologies (*i.e.*, radioisotope methods for determining assimilation efficiencies) has demonstrated, and explained mechanistically, that assimilation efficiencies vary widely among natural food types in invertebrates (Decho and Luoma 1991; Reinfelder and Fisher 1991). For example, assimilation efficiencies exceeding 20% occur for metals including silver, cadmium, chromium, cobalt, and zinc from at least some major types of food in molluscs (clams, mussels, oysters), whereas assimilation is lower from other food sources. Metals such as zinc, silver, selenium, and cadmium are assimilated from phytoplankton food sources by copepods (Reinfelder and Fisher 1991). In fresh water, bioaccumulation was found in rainbow trout and brown trout that were fed benthic fauna from a contaminated river (Mount *et al.* 1994; Woodward *et al.* 1994, 1995).

Models available to quantitatively predict bioaccumulation from multiple sources for a few species

These methodologies could be widely applied to invertebrate and probably fish species. Pharmacokinetic models have long been used to study metal bioaccumulation in fish, although such models have not been applied to the regulatory process. Recent studies have illustrated similar models for marine/estuarine bivalves (Thomann 1995; Wang *et al.* 1996; Luoma and Fisher 1997). Wang *et al.* (1996), experimentally derived the necessary physiological coefficients (assimilation efficiency from multiple food types, uptake rate from solution under different speciation regimes, efflux rates, ingestion rates) for a kinetic, pathway bioaccumulation model for mussels (*Mytilus edulis*). The relative contribution of food and water to bioaccumulation was predicted for the metal concentrations/ distributions, suspended solids loads, and salinities of San Francisco Bay and Long Island Sound. Approximately half of the bioaccumulation of silver, cadmium, and zinc originated from food in both ecosystems. Clams (*Macoma balthica*) take up a higher proportion of cadmium from food than mussels. Applying a similar model with more generic constants, Thomann (1995) also suggested that food was a predominant source of metal exposure in mussels. The tissue residues predicted from the laboratory-derived experiments agreed remarkably well with values observed in field studies and in earlier work on selenium by Luoma *et al.* (1992). Application of this approach to additional invertebrate species requires knowledge of at least the following:

1) Identification of food items ingested

2) Measurement of metal affinity for each food item

3) Measurement of the efficiency at which new species can assimilate metals associated with each food type

4) Determination of the dissolved uptake rate coefficients for the species

Links between toxicity and tissue residues derived from metal exposure via food or via multiple routes

Although studies have demonstrated linkages between the food exposure route and toxicity (Woodward *et al.* 1994, 1995 are recent examples), a critical unanswered question is whether toxicity emanating from exposure via food is additive or otherwise supplements toxicity emanating from dissolved exposures. Experimental difficulties in separating effects attributable to perturbed nutrition from those attributable to toxicants are examples of challenges in determining the contributions of metal-contaminated food to toxicity. Greater knowledge is needed about what mechanisms are responsible for the manifestation of toxicity originating from ingested metal. In some situations, like ecosystems with heavy contamination of sediments due to historical activities, dissolved metals may not adequately explain observed ecosystem toxicity (mortality, absences of metal-sensitive species, or sublethal impairment of individuals) (Farag *et al.* 1995; Luoma 1995). The food route of exposure should be considered as a cause of toxicity in such circumstances.

Strategies for linking tissue residues and toxicity

It is possible that whole body residues will correlate with toxicity in the cases of some organisms, and this empirical relationship can be used to derive criteria. However, complex detoxification mechanisms, among other processes, may confound such relationships. In some bivalves, the toxicity exerted by ingested metal is likely to be a function of a series of physiological processes, beginning with assimilation. Once internalized, metals can be depurated, sequestered at membranes (in insoluble granules or by metallothioneins), or remain in toxicologically active forms. Because sequestration may affect expression of toxicity, fractionation of tissues in ways that separate sequestered and soluble metals in tissues may aid in defining the bioactive tissue residue of metals. Alternatively, specific toxic mechanisms may be linked to residues in specific organs. However, where both ingestion and uptake of waterborne metal contribute to toxicity, it is unlikely that a single organ system, like the fish gill model, can be used to explain all adverse effects. Study of toxicity from multiple exposures is in its early stages, and it is questionable whether current knowledge allows immediate linkages.

Relationship between effects and time

A major issue in the application of laboratory toxicity data to natural systems is that exposures in the laboratory are to relatively constant concentrations for certain standard durations, whereas exposures in the field can have markedly different concentration-time courses. To address this issue, USEPA WQC are expressed in terms of mean concentration for averaging periods much shorter than the toxicity tests upon which the effect concentrations are based (*i.e.*, 1-h averages for the criterion maximum concentration [CMC] and 4-d averages for the criterion continuous concentration [CCC]). The intent of this approach is to preclude transiently high concentrations that could still elicit undesired effects even if the specified mean concentration is not exceeded. For example, for some chemicals, LC50s after an exposure of a few hours are not much higher than the 96-h LC50s used to set the USEPA CMC. For such chemicals, if the averaging period was set to 96 hours, concentrations could vary within this period enough for several hours to be lethal. This is especially true because the averaging period does not describe an isolated exposure, but rather a high exposure interval that is typically preceded and followed by additional significant exposure. Nevertheless, such short averaging periods are unduly restrictive for many chemicals and exposure conditions. A more accurate expression of risk, one that can relate laboratory toxicity results to widely varying exposure time series of concern in natural systems, is needed.

The effects of exposure time series on toxicity can be profound. Median lethal concentrations (LC50) for metals can be much higher for exposures of a few hours than they are for standard 48- or 96-h acute exposures (Bailey *et al.* 1985; Abel and Garner 1986). LC50 values often show an exponential decline with duration (Figure 4-4), approaching an asymptotic value referred to as the "incipient" or "threshold lethal concentration" (Sprague 1969). Depending on the organism and the chemical, the time required to approach the asymptote can vary from minutes to years. In practice, this idealized behavior can be perturbed by delayed response, acclimation, and complex mechanisms of toxicity, but an exponential decline will generally still be somewhat evident. Endpoints other than lethality can also be affected by exposure durations, and fluctuations in exposure concentrations can make responses even more complex. Studies on the effects of repeated short pulses of chemicals have generally shown greater effects than constant exposures on the basis of overall average exposure concentration (Brown *et al.* 1969; Ingersoll and Winner 1982; Seim *et al.* 1984; Siddens *et al.* 1986), the degree of this effect varying with the nature of the chemical, organism, endpoint, and exposure.

These effects of time on toxicity reflect the fact that a toxic response follows a chain of events—accumulation of toxicant at a site of action, disruption of biochemical processes at that site, and consequent impairment of physiological processes leading to the overt

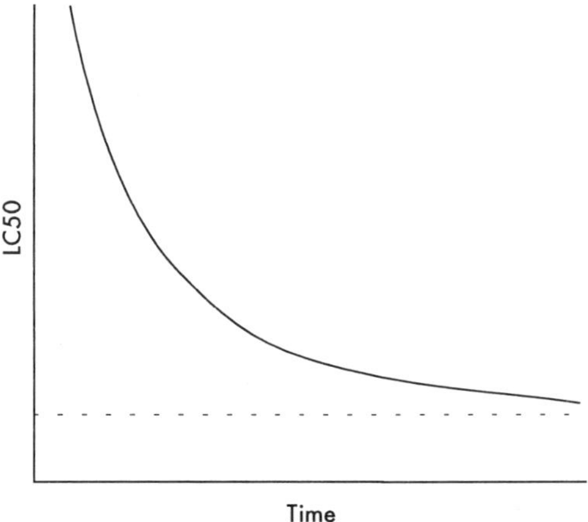

Figure 4-4   Idealized relationship of LC50 to exposure duration

toxic response—that each requires some time. Various authors (Mancini 1983; Neely 1984; Chew and Hamilton 1985) have noted that the exponential decline in LC50s is consistent with a model in which toxicity is elicited by accumulation of a chemical to a lethal threshold at a site of action. Reaching this threshold accumulation requires very high concentrations for short durations and progressively lower concentrations as increasing duration allows more time for uptake. Eventually, accumulation approaches steady state or equilibrium, so there is a minimum concentration needed for accumulation to reach the lethal threshold. These curves could also be considered to reflect "damage and repair" that encompass not only chemical accumulation but also the consequent effects on physiology. This explanation can be extended to other endpoints and to the effects of pulsed exposures.

Other authors have explicitly described the kinetics of toxicity in terms of the chemical accumulation needed to elicit effects, although most work has involved organics. For chlorinated benzenes, Van Hoogen and Opperhuizen (1988) not only reported that lethality was well correlated with attaining a threshold accumulation but also that differences among chemicals with respect to the magnitude of the LC50 and the speed of toxicity could be related to the toxicokinetics of the chemical. Some models include not only the kinetics of accumulation but also the kinetics of physiological damage necessary to elicit observed effects (Ankley *et al.* 1995). Models based on kinetics of accumulation can address both the effects of duration of constant exposures and of fluctuating exposures (Erickson *et al.* 1990; Hickie *et al.* 1995).

For metals, less success in this approach has been reported. Early attempts to relate fish kills to accumulation of metals in the gills suffered from a wide variability in metal accumulation data, and this has been a recurring problem. Nevertheless, with better methods for measuring, interpreting, and predicting accumulation at a site of action, models such as the gill-based approach ("receptor-loading model") have considerable promise. Such models provide a realistic mechanistic focus for describing the effects of metal speciation, hardness, and pH on metal accumulation and toxicity. In addition, they allow for better evaluations of hazard based on measurements of field-collected organisms. However, if they are intended to describe temporal aspects of toxicity, they must incorporate the kinetics as well as the extent of accumulation. Furthermore, because accumulation of metal at gills occurs quickly relative to the time needed to cause effects, the kinetics of the physiological responses to accumulation must also be incorporated if these models are to be used to predict the time course of toxicity.

Unfortunately, available knowledge of toxic mechanisms and toxicokinetics is often insufficient to support models that describe chemical accumulation and explicitly relate it to effects. However, the same conceptual model can be used to directly relate toxicity to exposure. As noted earlier, the general, idealized relationship of LC50s to exposure duration (Figure 4-4) reflects what would be expected based on accumulation of a chemical to a lethal threshold. Mancini (1983) discussed how such curves can be used to parameterize models equivalent to accumulation-based models and to predict the effects of any exposure time series. Such models can be considered to account for the kinetics of both accumulation and physiological response and can be extended to other endpoints (Connolly 1985), although some uncertainty can naturally arise from not explicitly describing individual processes.

Models based directly on the kinetics of toxic response have been the subject of several evaluations (Mancini 1983; Wang and Hanson 1985; Breck 1988; Hickie *et al.* 1995; Meyer *et al.* 1995). In general, the models provide fair approximations to the time course of toxicity in constant exposures and predict the general shift of effects between constant and pulsed exposures; however, they do not account for some responses. One extensive evaluation of these models has been made for acute toxicity of copper (Erickson 1985) and suggested that the models should be useful for applications to WQC, either by setting more appropriate averaging periods or by probabilistically predicting the extent of toxicity for expected exposures. With further analysis of available data, acute averaging periods longer than one hour could be demonstrated to be appropriate for metals criteria. Good parameterization of the models requires detailed response information that is not usually gathered in routine toxicity tests but that could be gathered with limited additional effort for many species of interest.

Water quality criteria and the protection of sediments

As we consider the development of alternative approaches for deriving WQC for the protection of water uses and aquatic life, it is important that adequate consideration is given to ensure that the approach also protects sediments and organisms that live in sediments. A number of approaches have been designed to derive SQC or sediment protective values (Chapman 1986; Barrick *et al.* 1988; Long and Morgan 1990; Di Toro *et al.* 1991; Long *et al.* 1995). The approach selected by the USEPA Office of Water for organics and metals is the equilibrium approach (Di Toro *et al.* 1991, 1992). The rationale for this approach is that carbon normalization (organics) and acid volatile sulfide (AVS) (metals) in conjunction with equilibrium theory can be used to provide a theoretical basis for calculating the bioavailable fraction of the chemical sorbed to the sediment. It provides a mechanistic approach to setting standards for the protection of sediments and benthic aquatic organisms. The procedures for deriving the sediment criteria are based upon the use of WQC, chemical-specific sediment partition coefficients, and AVS measurements. This approach appears to be based on sound scientific principles.

The development of the gill model for predicting the concentrations of metal in surface waters that are toxic to fish and potentially to other aquatic organisms opens the door to the possibility that this same approach may be used to predict toxicity from sediments to benthic organisms. Many benthic organisms possess a breathing apparatus that functions similarly to the gills of fish, or they adsorb oxygen and ions across surface membranes. In either case, it may be possible to use the ligand-binding concept to predict metal-induced toxicity from water for benthic organisms. This approach needs to be verified with experimental data. If proven to be scientifically valid, an assessment approach could be developed in conjunction with the equilibrium approach for deriving sediment criteria whereby the exposure route to the organism is calculated from the sediment via sediment interstitial water. Additional research would be needed to ensure that organisms which derive most of their residues from diet are also protected by this approach.

Recommendations

Recommendations for implementation and research actions are as follows.

Implementation

1) Water-effect ratios based on dissolved concentrations are a reasonable interim approach for site-specific modifications of metal standards, but they should eventually be replaced by methods and regulations that have a mechanistic basis. In the short term, the current use of the WER should be evaluated and the procedure refined as necessary.

2) Water regulation professionals should be educated in WER application, through both personal instruction and hands-on training.

3) Existing datasets on the time course of acute metal toxicity should be analyzed to provide recommendations regarding the duration of the time averaging period.

4) Acute-to-chronic ratios should continue as an interim method for setting standards protective of chronic endpoints, but they eventually should be replaced by methods based on a better understanding of the mechanisms underlying chronic toxicity and of the linkages between tissue residues and chronic toxicity.

5) Existing datasets relating toxicity to key water quality characteristics, including pH, hardness, and DOC, should be compiled and evaluated. These empirical, toxicity-based relationships can be used in the short term for improved imple-

mentation of metals criteria and in the long term for evaluation of mechanistic metal toxicity models.

Research

The following recommendations for research represent a structured framework for developing, testing, and validating the proposed approach to regulating metals in the environment:

1) Replacement of the existing approach with a tissue residue-related, mechanistically based toxicity approach (*e.g.*, "receptor-loading model") should be explored. The existing approach to metals regulation does not provide a rational basis for understanding toxicity and criteria derivation.

2) Research should be supported to pursue the gill modeling approach to acute toxicity as an alternative to the existing approach. The gill modeling approach may serve as an example for regulation of toxicity based on tissue residues. Specifically, there is a need for expanded research on gill residue versus mortality data for

   a) copper, cadmium, nickel, silver, lead, zinc, and chromium;

   b) multiple organisms, including invertebrates;

   c) metal mixtures;

   d) different water quality conditions;

   e) pulse exposure effects and time course;

   f) applicability to longer term exposures;

   g) mode of action, emphasizing clarification of mechanisms such as transport of $Ca2^{+}$ or Na^{+} and tissue accumulation of those ions; and

   h) kinetics of metal binding to gills.

3) Existing datasets should be used and new data should be developed to assess the relationship of acute toxicity to whole body residues, organ-specific residues, or tissue/organ/biochemical fractionations for key species of invertebrates.

4) Research is needed for expanding our knowledge of the mechanisms of metal toxicity in invertebrates and the linkage of these mechanisms to metal residues. This should include the effects of metals on $Ca+^{2}$ and Na^{+} transport, as shown in fish, and other mechanisms known to be affected by metals.

5) The next generation of criteria for protecting aquatic ecosystems should consider multiple routes of exposures (*e.g.*, dietary as well as waterborne). Sediments and water should not be considered separately. There should be a melding of water and sediment criteria considering the relationships of toxicity to tissue residues, routes of exposure, and fate.

6) For chronic toxicity, establishing linkages between specific tissue residues (*e.g.*, gill or liver residues in fish) and mechanisms of toxicity should be a high

priority in the development of the tissue-residue, mechanism-based approach to regulation.

7) Ecosystem response should be included in evaluating chronic toxicity. It is essential to identify which processes and effects on individuals determine how populations and communities respond.

References

Abel PD, Garner SM. 1986. Comparisons of median survival times and median lethal exposure times for *Gammarus pulex* exposed to cadmium, permethrin, and cyanide. *Water Res* 20:579–582.

Allison JD, Brown DS, Novo-Gradac KJ. 1991. MINTEQA2/PRODEFA2, a geochemical assessment model for environmental systems: Version 3.0 User's manual. Washington DC: USEPA.

Ankley GT, Erickson RJ, Phipps GL, Mattson VR, Kosian PA, Sheedy BR, Cox JS. 1995. Effects of light intensity on the phototoxicity of fluoranthene to a benthic macroinvertebrate. *Environ Sci Technol* 29:2828–2833.

Bailey HC, Liu DHW, Javitz HA. 1985. Time/toxicity relationships in short-term static, dynamic, and plug-flow bioassays. In: Bahner RC, Hansen, DJ, editors. Aquatic toxicology and hazard assessment: Eight symposium. Philadelphia PA: American Soc for Testing and Materials (ASTM). STP 891. p 193-212.

Barrick RC, Becker DS, Brown L, Beller H, Pastorok RA. 1988. Sediment quality values refinement: 1988 update and evaluation of Puget Sound AET. Bellevue WA: PTI Environmental Services. PTI Contract C717-02.

Borgmann U. 1981. Determination of free metal ion concentration using bioassays. *Can J Fish Aquat Sci* 38:999–1002.

Breck JE. 1988. Relationships among models for acute toxic effects: applications to fluctuating concentrations. *Environ Toxicol Chem* 7(9):775–778.

Brown VM, Jordan DHM, Tiller BA. 1969. The acute toxicity to rainbow trout of fluctuating concentrations and mixtures of ammonia, phenol and zinc. *J Fish Biol* 1:1–9.

Carlson AR, Brungs WA, Chapman GA, Hansen DJ. 1984. Guidelines for deriving numerical water quality criteria by modifying national criteria. Duluth MN: USEPA Office of Research and Development, Environmental Research Laboratory. EPA/600/3-84/099.

Chapman PM. 1986. Sediment quality criteria from the sediment quality triad: an example. *Environ Toxicol Chem* 5(11):957–964.

Chew RD, Hamilton MA. 1985. Toxicity curve estimation: fitting a compartment model to median survival times. *Trans Am Fish Soc* 114:403–412.

Connolly JP. 1985. Predicting single-species toxicity in natural water systems. *Environ Toxicol Chem* 4(4):573–582.

Daly HR, Campbell IC, Hart BT. 1990a. Copper toxicity to *Paratya australiensis:* I. Influence of nitrilotriactetic acid and glycine. *Environ Toxicol Chem* 9(8):997–1006.

Daly HR, Campbell IC, Hart BT. 1990b. Copper toxicity to *Paratya australiensis:* II. Influence of bicarbonate and ionic strength. *Environ Toxicol Chem* 9(8):1007–1011.

Daly HR, Hart BT, Campbell IC. 1992. Copper toxicity to *Paratya australiensis:* IV. Relationship with ecdysis. *Environ Toxicol Chem* 11(6):881–883.

Daly HR, Jones MJ, Hart BT, Campbell IC. 1990c. Copper toxicity to *Paratya australiensis:* III. Influence of dissolved organic matter. *Environ Toxicol Chem* 9(8):1013–1018.

Decho AW, Luoma SN. 1991. Time-courses in the retention of food material in the bivalves *Potamocorbula amurensis* and *Macoma balthica*: significance to the assimilation of carbon and chromium. *Mar Ecol Prog Ser* 78:303–314.

Di Toro DM, Mahoney JD, Hansen DJ, Scott KJ, Carlson AR, Ankley GT. 1992. Acid volatile sulfide predicts the acute toxicity of cadmium and nickel in sediments. *Environ Sci Technol* 26:96-101.

Di Toro DM, Zarba CS, Hansen DJ, Berry WJ, Swartz RC, Cowan CE, Pavlou SP, Allen HE, Thomas NA, Paquin PR. 1991. Technical basis for establishing sediment quality criteria for nonionic organic chemicals using equilibrium partitioning. *Environ Toxicol Chem* 10(12):1541–1583.

Elseroad HJ, Rue Jr WJ, Hinckley DA, Woodis HL. 1994. Chemical translator guidance manual. Palo Alto CA: Electric Power Research Institute. EPRI Report #TR-104047.

Erickson R, Kleiner C, Fiandt J, Highland T. 1990. Use of toxicity models to reduce uncertainty in aquatic hazard assessments: effects of exposure conditions on pentachlorethane toxicity to fathead minnows. Duluth MN: USEPA Report.

Erickson RJ. 1985. Evaluation of a model for the prediction of the effect of fluctuating concentrations on the lethality of copper to fathead minnows. Duluth MN: USEPA Report.

Evans DH. 1987. The fish gill: site of action and model for toxic effects of environmental pollutants. *Environ Health Perspect* 71:47-58.

Farag AM, Stansbury MA, Hogstrand C, MacConnell E, Bergman HL. 1995. The physiological impairment of free-ranging brown trout exposed to metals in the Clark Fork River, Montana. *Can J Fish Aquat Sci* 52:2038–2050.

Hickie BE, McCarty LS, Dixon DG. 1995. A residue-based toxicokinetic model for pulse-exposure toxicity in aquatic systems. *Environ Toxicol Chem* 14(12):2187–2197.

Hogstrand C, Wilson RW, Polgar D, Wood CM. 1994. Effects of zinc on the kinetics of branchial calcium uptake in freshwater rainbow trout during adaptation to waterborne zinc. *J Exp Biol* 186:55–73.

Ingersoll CG, Winner RW. 1982. Effect on *Daphnia pulex* (De Geer) of daily pulse exposures to copper or cadmium. *Environ Toxicol Chem* 1(4):321–327.

Janes N, Playle RC. 1995. Modeling silver binding to gills of rainbow trout (*Oncorhynchus mykiss*). *Environ Toxicol Chem* 14(11):1847–1858.

Laurén DJ, McDonald DG. 1986. Influence of water hardness, pH, and alkalinity on the mechanisms of copper toxicity in juvenile rainbow trout, *Salmo gairdneri. Can J Fish Aquat Sci* 43:1488-1496.

Laurén DJ, McDonald DG. 1987. Acclimation to copper by rainbow trout, *Salmo gairdneri:* physiology. *Can J Fish Aquat Sci* 44:99–104.

Lin H, Randall DJ. 1990. The effect of varying water pH on the acidification of expired water in rainbow trout. *J Exp Biol* 149:149-1 60.

Long ER, MacDonald DD, Smith SL, Calder FD. 1995. Incidence of adverse biological effects within ranges of chemical concentrations in marine and estuarine sediments. *Environ Manage* 19(1):81–97.

Long ER, Morgan LG. 1990. The potential for biological effects of sediment-sorbed contaminants tested in the National Status and Trends Program. Seattle WA: Office of Oceanography and Marine Assessment. NOAA Technical Memorandum NOS OMA 52.

Luoma SN. 1995. Prediction of metal toxicity in nature from bioassays: limitation and research needs. In: Tessier A, Turner D, editors. Metal speciation and bioavailability in aquatic systems. Sussex, England: J Wiley. p 610–659.

Luoma SN, Fisher NS. 1997. Uncertainties in assessing contaminant exposure from sediments. In: Biddinger GR, Dillon T, Ingersoll CG, editors. Ecological risk assessment of contaminated sediments. Pensacola FL: SETAC Pr.

Luoma SN, Johns C, Fisher NS, Steinberg NA, Oremland RS, Reinfelder J. 1992. Determination of selenium bioavailability to a benthic bivalve from particulate and solute pathways. *Environ Sci Technol* 26:485–497.

MacRae RK, Smith DE, Swoboda-Colberg N, Meyer JS, Bergman HL. 1997. Copper binding affinity of rainbow trout (*Oncorhynchus mykiss*) and brook trout (*Salvelinus fontinalis*) gills. *Environ Toxicol Chem* (in press).

Mallat J. 1985. Fish gill structural changes induced by toxicants and other irritants: a statistical review. *Can J Fish Aquat Sci* 42:630-648.

Mancini JL. 1983. A method for calculating effects on aquatic organisms of time-varying concentrations. *Water Res* 17:1355–1361.

McDonald DG, Wood CM. 1993. Branchial mechanisms of acclimation to metals in freshwater fish. In: Rankin JC, Jensen FB, editors. Fish ecophysiology. London: Chapman and Hall. p 297-321.

Meyer JS, Gulley DD, Goodrich MS, Szmania DC, Brooks AS. 1995. Modeling toxicity due to intermittent exposure of rainbow trout and common shiners to monochloramine. *Environ Toxicol Chem* 14(1):165–175.

Morel FMM. 1983. Principles of aquatic chemistry. Toronto, ON: J Wiley. 446 p.

Morel FMM, Hering JG. 1993. Principles and applications of aquatic chemistry. New York: J Wiley. 588 p.

Mount DR, Barth AK, Garrison TD, Barten KA, Hockett JR. 1994. Dietary and waterborne exposure of a rainbow trout (*Oncorhynchus mykiss*) to copper, cadmium, lead and zinc using a live diet. *Environ Toxicol Chem* 13(12):2031–2041.

Neely WB. 1984. An analysis of aquatic toxicity data: water solubility and acute LC50 fish data. *Chemosphere* 7:813–819.

Pagenkopf GK. 1983. Gill surface interaction model for trace-metal toxicity to fishes: role of complexation, pH, and water hardness. *Environ Sci Technol* 17:342-347.

Pilgaard L, Malte H, Jensen FB. 1994. Physiological effects and tissue accumulation of copper in freshwater rainbow trout (*Oncorhynchus mykiss*) under normoxic and hypoxic conditions. *Aquat Toxicol* 29:197–212.

Playle RC, Dixon DG, Burnison K. 1993a. Copper and cadmium binding to fish gills: estimates of metal-gill stability constants and modelling of metal accumulation. *Can J Fish Aquat Sci* 50:2678-2687.

Playle RC, Dixon DG, Burnison K. 1993b. Copper and cadmium binding to fish gills: modification by dissolved organic carbon and synthetic ligands. *Can J Fish Aquat Sci* 50:2667–2677.

Playle RC, Goss GG, Wood CM. 1989. Physiological disturbances in rainbow trout (*Salmo gairdneri*) during acid and aluminum exposures in soft waters of two calcium concentrations. *Can J Zool* 67:314-324.

Playle RC, Wood CM. 1989. Water pH and aluminum chemistry in the gill micro-environment of rainbow trout during acid and aluminum exposures. *J Comp Physiol B Metab Transp Funct* 159:539–552.

Prothro MG. 1993. Office of water policy and technical guidance on interpretation and implementation of aquatic metals criteria. Memorandum from Acting Assistant Administrator for Water. Washington DC: USEPA Office of Water. 7 p. Attachments 41 p.

Reinfelder JR, Fisher NS. 1991. The assimilation of elements ingested by marine copepods. *Science* 251:794–796.

Richards J, Kuehn A, Hughes C, Playle RC. Unpublished. Wilfrid Laurier University, Department of Biology. Waterloo, Ontario, Canada.

Schecher WD, McAvoy DC. 1992. MINEQL+: a software environment for chemical equilibrium modeling. *Computers, Environment and Urban Systems* 16:65–76.

Seim WK, Curtis LR, Glenn SW, Chapman GA. 1984. Growth and survival of developing steelhead trout continuously or intermittently exposed to copper. *Can J Fish Aquat Sci* 41:433–483.

Siddens LK, Seim WK, Curtis LR, Chapman GA. 1986. Comparison of continuous and episodic exposure to acidic, aluminum-contaminated waters of brook trout (*Salvelinus fontinalis*). *Can J Fish Aquat Sci* 43:2036–2040.

Skidmore JF, Tovell PWA. 1972. Toxic effects of zinc sulphate on the gills of rainbow trout. *Water Res* 6:217-230.

Sprague JB. 1969. Measurement of pollutant toxicity to fish: I. Bioassay methods for acute toxicity. *Water Res* 3:793–821.

Spry DJ, Wood CM. 1985. Ion flux rates, acid-base status, and blood gases in rainbow trout, *Salmo gairdneri*, exposed to toxic zinc in natural softwater. *Can J Fish Aquat Sci* 42:1332–1341.

Staurnes M, Sigholt T, Reite OB. 1984. Reduced carbonic anhydrase and Na+, K$^+$ ATPase activity in gills of salmonids exposed to aluminum-containing acid water. *Experientia* 40:226–227.

Stephan CE, Mount DI, Hansen DJ, Gentile JH, Chapman GA, Brungs WA. 1985. Guidelines for deriving numerical water quality criteria for the protection of aquatic organisms and their uses. USEPA Office of Research and Development. PB85-227049.

Sunda WG, Guillard RRL. 1976. The relationship between cupric ion activity and the toxicity of copper to phytoplankton. *J Mar Res* 34:511–529.

Thomann RV, Mahoney JD, Mueller R. 1995. Steady-state model of biota sediment accumulation factor for metals in two marine bivalves. *Environ Toxicol Chem* 14(11):1989–1998.

[USEPA] U.S. Environmental Protection Agency. 1983. Water quality standards handbook. Washington DC: Office of Water, Regulations and Standards.

[USEPA] U.S. Environmental Protection Agency. 1984. Guidelines for deriving numerical aquatic site-specific water quality criteria for modifying national criteria. Duluth MN: Environmental Research Laboratory. EPA/600/3-84/099 or PB85-121101.

[USEPA] U.S. Environmental Protection Agency. 1986. Quality criteria for water 1986. Washington DC: Office of Water, Regulations and Standards. EPA/440/5-86/001.

[USEPA] U.S. Environmental Protection Agency. 1992. Interim guidance on interpretation and implementation of aquatic life criteria for metals. Washington DC: Office of Science and Technology, Health and Ecological Criteria Division.

[USEPA] U.S. Environmental Protection Agency. 1994. Interim guidance on determination and use of water-effect ratios for metals. Washington DC: Office of Water, Office of Science and Technology. EPA/823/B-94/001. 154 p.

[USEPA] U.S. Environmental Protection Agency. 1995. Water quality standards; establishment of numeric criteria for priority toxic pollutants; states compliance revision of metals criteria (Action: Interim final rule, notice of data availability and request for comments). *Federal Register*, 40 CFR Part 131. May 4, 1994. p 22229–22237.

[USEPA] U.S. Environmental Protection Agency. 1996. The metals translator: guidance for calculating a total recoverable permit limit from a dissolved criterion. Kinerson RS, Mattice JS, Stine JF, editors. Washington DC: Office of Water. EPA 823-B-96-007. 60 p.

Van Hoogen G, Opperhuizen A. 1988. Toxicokinetics of chlorobenzenes in fish. *Environ Toxicol Chem* 7(3):213–219.

Verbost PM, Flik G, Locke RAC, Wendelaar Bonga SE. 1987. Cadmium inhibition of Ca^{2+} uptake in rainbow trout gills. *Am J Physiol* 253:R216-R221.

Verbost PM, Van Rooij J, Flik G, Lock RAC, Wendelaar Bonga SE. 1989. The movement of cadmium through freshwater trout branchial epithelium and its interference with calcium transport. *J Exp Biol* 145:185–197.

Wang MP, Hanson SA. 1985. The acute toxicity of chlorine on freshwater organisms: time-concentration relationships of constant and intermittent exposures. In: Bahner RC, Hansen DJ, editors. Aquatic toxicology and hazard assessment: Eighth symposium. Philadelphia PA: American Soc for Testing and Materials (ASTM). STP 891. p 213–232.

Wang W-X, Fisher NS, Luoma SN. 1996. Kinetic determination of trace element bioaccumulation in the mussel *Mytilus edulis. Mar Ecol Prog Ser* (in press).

Welsh PG, Skidmore JF, Spry DJ, Dixon DG, Hodson PV, Hutchinson NJ, Hickie BE. 1993. Effect of pH and dissolved organic carbon on the toxicity of copper to larval fathead minnow (*Pimephales promelas*) in natural lake waters of low alkalinity. *Can J Fish Aquat Sci* 50:1356–1362.

Wilson RW, Taylor EW. 1993. The physiological responses of freshwater rainbow trout, *Oncorhynchus mykiss*, during acutely lethal copper exposure. *J Comp Physiol B Metab Transp Funct* 163:38–47.

Wood CM. 1992. Flux measurements as indices of H+ and metal effects on freshwater fish. *Aquat Toxicol* 22:239–264.

Wood CM, Hogstrand C, Galvez F, Munger RS. 1996. The physiology of waterborne silver toxicity in freshwater rainbow trout (*Oncorhynchus mykiss*): I. The effects of ionic Ag+. *Aquat Toxicol* 35:93–109.

Woodward DF, Brumbaugh WG, DeLonay AJ, Little EE, Smith CE. 1994. Effects on rainbow trout fry of a metals-contaminated diet of benthic invertebrates from the Clark Fork River, Montana. *Trans Am Fish Soc* 123:51–62.

Woodward DF, Farag AM, Bergman HL, DeLonay AJ, Little EE, Smith CE, Barrows FT. 1995. Metals-contaminated benthic invertebrates in the Clark Fork River, Montana: effects on age 0 brown trout and rainbow trout. *Can J Fish Aquat Sci* 52:1994–2004.

Xue H, Sigg L. 1993. Free cupric ion concentration and Cu(II) speciation in a eutrophic lake. *Limnol Oceanogr* 38:1200–1213.

Chapter 5

Chemical speciation and metal toxicity in surface freshwaters

James R. Kramer, chair
Herbert E. Allen, William Davison, Kathy L. Godtfredsen,
Joseph S. Meyer, E. Michael Perdue, Edward Tipping, Dik van de Meent, John C. Westall

The Environmental Chemistry Workgroup considered current knowledge of metal speciation and how it could be used in the assessment of aquatic metal toxicity. Only oxic freshwaters are considered, since sufficient data exist to assess and to model the oxic freshwaters system reasonably well. Issues of modeling, measurement, and relationship of metal speciation to biological effects are considered.

Figure 5-1 is a schematic for a hypothetical system involving aqueous, biological, and sediment compartments. The dashed enclosure considers that portion of the overall system considered here. A metal, M, is assumed to be in equilibrium with dissolved ligands (ML^{m-n}) and bound to suspended particulate matter ($M \equiv S_p$), biomass ($M \equiv S_b$), and sediment ($M \equiv S_s$). The total metal, ΣM, is the sum of the various forms and is distributed among the various forms dependent upon the binding constants (distribution coefficients).

Equilibrium thermodynamic modeling of dissolved metal speciation in oxic freshwaters is assumed in this assessment. If an equilibrium model is used to predict the uptake or toxicity of a metal ion, then rapid and reversible equilibrium among the various reactants and products is mandated. Assuming equilibrium, any perturbation will result in redistribution of M, dependent upon the binding constants. For this assumption, on the one hand, all metal forms can be considered "bioavailable." On the other hand, one may consider bioavailability as the specific chemical reaction and form of the metal associated with a biological receptor (see Chapter 4, section "Generation of metals criteria using a receptor loading model: a new approach"). For example, only aquo Cu^{2+} may be assumed to react and affect a biological binding site. If copper exists in solution primarily as a complex, however, it might be considered "not bioavailable." The two usages of bioavailability converge if the metal system is well buffered. The metal ion, M^{m+}, in the aqueous phase, will be well buffered for a given total aqueous metal concentration if the reactions involving M^{m+} are reversible and if the ligand, L^{n-}, is large compared to M^{m+}. In the real world, it is important to determine what portions of the system are in equilibrium.

A minimum list of master variables needed to describe metal speciation for oxic freshwater systems includes pH, alkalinity, hardness (preferably Ca^{2+} concentration), ionic strength (total dissolved solids [TDS]), DOC, total dissolved metal, unique anthropogenic inputs (_e.g.,_ EDTA), and preferably Al and Fe(III). The concentrations of TSS and POM, along with trace metal concentrations, are needed to describe suspended particulate concentration. Generally, DOC appreciably affects metal speciation.

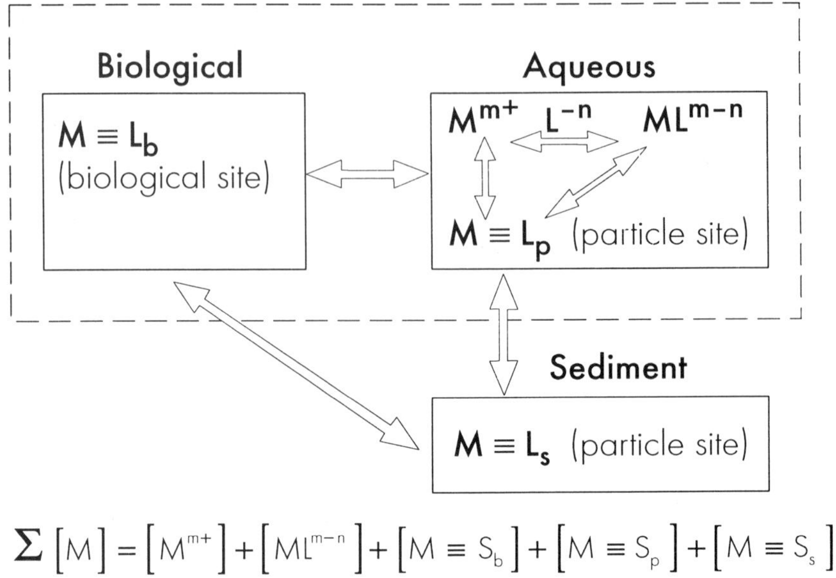

$$\sum [M] = [M^{m+}] + [ML^{m-n}] + [M \equiv S_b] + [M \equiv S_p] + [M \equiv S_s]$$

Figure 5-1   Schematic of a reversible aquatic system, considering aquatic, biological, and sediment compartments

Analyses of soluble and particulate metal are operationally defined. Discussion on the manner in which soluble and dissolved metal concentrations are estimated by the USEPA is found in Chapter 3, section "Description of current regulatory processes." Additional discussion is found in the section titled "The 'dissolved' fraction" in this chapter.

In general, aquatic systems can be separated into categories "hard" and "soft." These waters are characterized by differences in pH, alkalinity, hardness, and TDS, and this water classification is used in assessments of model results.

Suspended particulate matter represents important and complex surfaces for metal uptake and release. These particles include minerals, biota, and organic macro molecules, ranging in size from 0.001 to 10 µm (Buffle and Leppard 1995). These diverse suspended particles are very important in metal uptake/release and transport because of their large surface areas, strong binding constants, and sorptive capacities.

Although DOC may dominate metal binding in many aquatic environments, including POTW outfalls and receiving waters, suspended solids and reduced sulfur species may control metal binding in low-pH aquatic environments such as those found in mining areas and in elevated TDS (high sulfate) environments. The trace metal speciation description given here may apply if these systems are routed through wetlands.

The discussion that follows focuses on the speciation in solution. Then linking solution speciation to the gill model ($M \equiv S_b$), presented in Chapter 4, is discussed.

State of knowledge regarding solution speciation modeling

There are a number of aqueous speciation models extant. Discussion of these models ensues with the conclusion that Model V of WHAM (Tipping 1994) is the more appropriate model to use. An error analysis is then carried out for WHAM, and requirements of kind and quality of data are considered.

Comparison of equilibrium models

Chemical speciation models have reached a sufficient level of maturity that the reactions of very complex mixtures of ligands such as dissolved organic material (DOM) with metal ions can be modeled reasonably well. The current generation of models, *i.e.*, MINTEQA2 (Allison and Purdue 1994), MINEQL$^+$ (Schecher and McAvoy 1995), and WHAM (Tipping 1994) use a variety of conceptual schemes to simulate chemical heterogeneity, to handle electrostatic effects, and to treat competitive binding of cations. The similarities between the elements of current models outweigh their differences.

The ultimate success of a chemical speciation model depends on the quality of its thermodynamic database, and in this regard, current models differ significantly. MINTEQA2, which was developed and distributed by the USEPA, contains an inorganic database, a solids database, an organic ligand database for simple organic ligands like EDTA and citrate, and a "composite ligand" database for DOM. The inorganic and solids databases are believed to be of generally high quality; however, the simple organic ligand database contains erroneous entries, especially for metal complexes with protonated ligands (Serkiz *et al.* 1996). The composite ligand database is optimized for proton and metal binding by DOM from the Suwannee River, and the metal binding parameters all come from studies of the competitive binding of europium(III) and other metal ions, generally at low pH and high metal ion concentrations.

In contrast, the WHAM database has been optimized for metal binding for 30 metal ions, using over 80 published datasets for a variety of soil fulvic acids and DOM samples, and over a pH range of 2 to 9. Furthermore, WHAM and its forerunners have been shown to account reasonably well for available field data on metal speciation (Tipping *et al.* 1991).

MINTEQA2 and MINEQL$^+$ have good and extensive databases for inorganic solutes and solids, a very broad user base, and a fairly complete set of optional subroutines for treating ion exchange, sorption, and partitioning equilibria. We recommend that Model V and its database for DOM be incorporated into MINTEQA2 and MINEQL$^+$.

Applying WHAM to surface waters

Chemical input data required to run the model for metal speciation are pH and organic matter, major cations and anions, and the concentration of potentially toxic metals of interest. In the absence of a complete dataset, pH and DOC are mandatory, and useful results can be obtained by making reasonable estimates of other concentrations, *e.g.*, on the basis of conductivity and hardness. In some cases, additional data may be required, *i.e.*, sulfide ligand concentrations in the case of Ag or specific organic ligands known to be components of a particular waste.

The concentration of DOC provides an estimate of the concentration of complexing organic matter. Dissolved organic material is obtained from DOC on the basis of carbon content (approximately 50%). In practice, it has been found that metal complexation is overestimated if all of the DOM is taken to be fulvic acid, and a correction factor has to be applied. The current best estimate of this factor is 0.65, based on application of WHAM to various field data, and this value is adopted here.

How good are speciation calculations?

There are three sources of error in calculated speciation, even assuming that equilibrium exists: model limitations, data quality, and parameter uncertainty.

Model limitations. Potential limitations include extrapolating the model to situations for which it has been insufficiently parameterized. Laboratory data do not represent higher pH conditions (>7) very well, and most refer to situations where metal concentrations, bound and free, are higher than in natural waters. Differences in the properties of different DOM would also manifest themselves. It is difficult to estimate the significance of DOM variability, however, because most data have been generated on isolated samples whose heterogeneity may be less than that of whole DOM (Shuman 1990). For the time being, we are obliged to assume that, with respect to metal binding, DOM in all natural waters is the same. There is a need for more field testing in order to evaluate this assumption and, if necessary, to take any variability in DOM properties into account.

Data quality. The extent to which bad data affect model predictions varies from one water quality variable to another. As an example, a simple exercise was conducted at this workshop. For hypothetical soft and hard waters having different concentrations of DOC, dissolved inorganic carbon (DIC), Ca^{2+}, Na^+, $Cl-$, and different pH values, the effects of analytical errors on the predicted chemical speciation of copper(II) and cadmium(II) were examined. Errors of ± 0.2 pH and $\pm 10\%$ in DOC and DIC concentrations were imposed so as to either maximize or minimize the predicted concentration of the free metal ion. The results are summarized in Table 5-1. It is seen that for both waters the errors resulted in calculated $[Cu^{2+}]$ differences of a factor of about 2, but $[Cd^{2+}]$ changed by less than 25%. Further calculations revealed that most of the variability in predicted free metal ion concentrations is caused by pH errors. The next most important parameter is the DOC concentration. It is interesting to note that, whereas for copper, the concentration of free aquo ion is lower in the higher pH hard water, the reverse is found for cadmium. This reflects the fact that Ca^{2+}, the concentration of which is higher in the hard water, competes more effectively with cadmium than with copper for DOM binding sites.

Parameter uncertainty. Published metal binding data were analyzed in deriving parameter values for Model V, which in WHAM deals with organic complexation (Tipping and Hurley 1992; Tipping 1993a, 1993b). The key parameter with respect to metal binding is pK_{MHA}, the negative logarithm of the intrinsic equilibrium constant for metal-proton exchange. When several datasets for a given metal were analyzed, it was found

Table 5-1 Effects of data error on calculated speciation of copper and cadmium in typical soft and hard waters

Variable	Soft water			Hard water		
	Default	+ Bias	- Bias	Default	+ Bias	- Bias
pH	6	5.8	6.2	8	7.8	8.2
DOC, mg/L	15	13.5	16.5	5	4.5	5.5
DIC, mM	0.5	0.45	0.55	2	1.8	2.2
Ca^{2+}, mM	0.1	0.11	0.09	1	1.1	0.9
$[Cu^{2+}]$, μM	0.0044	0.0087	0.0022	0.00059	0.0011	0.00031
$[Cd^{2+}]$, μM	0.31	0.38	0.24	0.55	0.64	0.46

In both cases, total dissolved metal concentration was taken to be 1 μM. Concentration of NaCl was 0.1 mM in the soft water and 1mM in the hard water. The "+ bias" heading indicates that errors were imposed in order to increase the concentration of free aquo ion; "- bias" indicates decrease in the concentration.

that although the values of pK_{MHA} were similar, there were differences. To obtain a "best average" default value, the mean pK_{MHA} was taken. Typically, the range of values varied by ± 0.2 log units. The error in actual binding sites is amplified, however, because the model generates bidentate sites by combining equilibrium constants for monodentate sites. To investigate what effect errors of data inaccuracy or parameter uncertainty might have on speciation predictions, trial calculations with WHAM were performed. The results are presented in Table 5-2. It is found that for copper, order-of-magnitude variation in Cu^{2+} concentration occurs, but for cadmium the ranges are smaller. In general, it is expected that when a significant fraction of the metal ion is present as the free ion, errors in its predicted concentration are relatively insensitive to errors in pK_{MHA} (Table 5-2). In contrast, if the free metal ion concentration is very low, then its predicted concentration is more sensitive to these data quality errors.

A more complete analysis of model sensitivity to data and parameter uncertainty would require a series of Monte Carlo simulations, in which model predictions are generated from distributions of water quality data (see *e.g.*, Tipping *et al.* 1990).

Modeling partitioning to particles

Models for sorption to particulates were also considered to determine if they could be incorporated in near-term regulatory efforts. The analysis is presented in this section.

Comparison of equilibrium and surface complexation models

Partitioning of metals between water and solids occurs through sorption and precipitation. Precipitation of metal salts can occur if the solubility product, K_s, for the reaction is exceeded. Presently, (equilibrium) speciation models are able to predict oversaturation of sparingly soluble metal salts. However, the actual conditions under which precipitation occurs, and the rates of precipitation and dissolution, do not seem to be understood well

Table 5-2 Effects of parameter uncertainty on WHAM-calculated concentrations of free aquo ion

	Soft			Hard		
	pK_{MHA} def	$+pK_{MHA}$ 0.2	$-pK_{MHA}$ 0.2	pK_{MHA} def	$+pK_{MHA}$ 0.2	$-pK_{MHA}$ 0.2
$[Cu^{2+}]$, µM	0.0044	0.026	0.00054	0.00059	0.0040	0.000064
$[Cd^{2+}]$, µM	0.31	0.026	0.17	0.55	0.83	0.17

Water compositions are as for the "normal" soft and hard waters as shown in Table 5-1.

enough to be incorporated into predictive metal partitioning models. Progress has been made recently in understanding the chemistry of metal sulfides under anoxic conditions (Luther *et al.* 1996).

Laboratory studies of metal binding to particles have concentrated on describing the exchange of metals with protons on well-characterized oxide surfaces. Various surface complexation models can generally describe the observed dependency of the binding on pH and ionic strength in these simple systems. Proton exchange reactions and ionic strength effects, using surface charge relationships, are associated with pH effects. When such models are combined with speciation codes, the complete surface and solution speciation can be described. For a well-defined surface, such as hydrous ferric oxide (Dzombak and Morel 1990), reasonable agreement on a set of binding constants for protons and metal ions is emerging, allowing the prediction of the partition of metal between solution and solid phase for a range of solution compositions.

Natural particles are mixtures of mineral phases, biotic and abiotic organic particles, and hydrous metal oxides, the proportions and characteristics of which are rarely well described. Because their compositions and surface chemistries are poorly defined, the surface complexation approach cannot be readily used for predicting the partition between solution and solid phase. Moreover, whereas the binding characteristics of DOM can be regarded as universal and therefore can be incorporated into a general predictive model such as Model V, the inherent variability in particulate material prevents its characterization by a single set of capacities and binding constants.

Nevertheless, pH-dependent partition models may be used to predict site-to-site variation in partition coefficients within one water system. For soils, there is empirical evidence to demonstrate that the major controls of solids-water partitioning are pH and organic matter (Lee *et al.* 1996).

From a practical point of view and for the purpose of WLA, it is necessary to measure empirical site-specific partition coefficients for suspended solids. State-of-the-art science in modeling sorption in the general case is not developed enough for incorporation into regulations.

Recommended approach
Because modeling partitioning to natural particles within natural waters has not reached a state of practical application, the USEPA model for the metals translator (USEPA 1996)

was reviewed from a chemical perspective. The purpose of the review was to assess consistency between the recommended collection and analysis techniques and parameters from the literature and the techniques and parameters proposed as part of the regulatory process. The metals translator is defined as the fraction of total recoverable metal in the effluent (C_T) that is dissolved in the receiving water (C_D). In general, the translator method is recommended because it is a practical, straightforward, empirical approach.

As acknowledged in the USEPA (1996) guidance, the relationship between the C_T and C_D is dependent upon a number of variables and is expected to vary significantly both temporally and spatially. For this reason, it is highly recommended that the minimum dataset required in the guidance be expanded to ensure adequate understanding of the range of conditions likely at a site before the site-specific metal translator field approach is finalized. Included in this minimum dataset should be a range of pH, TSS, hardness, conductivity or salinity, DOC concentrations, and the concentrations of regulated metals.

In addition, because these studies will be planned, executed, and interpreted by a diverse group of individuals, it is recommended that all individuals who will be involved in each study (*i.e.*, the regulated community and all involved regulators) participate in early and intensive discussion to define site-specific protocols, including location of sampling, frequency of sampling, and statistical interpretation of data.

A possible improvement on the approach is to use a more fundamentally based, but still simple, expression for K_D, in which the amount of adsorbed metal is explicitly related to activities of free metal and hydrogen ion. This has been proposed by Mouvet and Bourg (1983). We write

$$K_D^* = X_M^* \frac{a_H}{a_{M^{z+}}}$$

where X_M is the moles of metal adsorbed per gram of solids, and "a" represents activity. The value of a_M would be obtained from a speciation calculation of dissolved metal, and a_H would be obtained from pH. This simple approach would tend to make variations in K_D depend mostly on variations in solids properties, and it would tend to remove dependence on solution chemistry. As a result, K_D^* might have a smaller or more readily definable variability, both temporally at a given site and among sites, than does the current expression for K_D.

Issues of measurement

The acquisition of useful and meaningful analytical data was considered as a key component of effective metals regulations. This section assesses several critical measurement issues.

The "dissolved" fraction

Current practice is to quantify the metal in a water by measuring one or more of the following fractions: total, total recoverable, acid soluble, or dissolved. The recommended procedure for analysis of dissolved metals involves filtration through a 0.45 μm membrane. Such a procedure results in a filtrate that includes colloidal fractions. If the colloids comprise organic material to which metals are sorbed, the speciation modeling is relatively unaffected because this material is included in the similarly operationally defined (0.45 µm) DOM. However, if the colloids combine other materials that bind metals, such as hydrous metal oxides, the dissolved fraction to be used in the speciation code is artificially high. The problem of overestimating the dissolved fraction can be partly overcome by using a smaller pore-sized filter, but there is no consensus regarding the best size, and other errors may be introduced if very small pore sizes are used.

Procedures for directly measuring well-defined species based on, for example, electrochemical techniques (Tercier and Buffle 1993) or gel membranes (Davison and Zhang 1996) in the unfiltered solution are currently being developed. When available, they will be useful in confirming the speciation model results in laboratory solutions and for validating the models in field conditions. Until such procedures are adequately validated, filtration procedures should continue to be used.

Importance of measurement technique standardization
The success of any model prediction is directly dependent on the quality of input data. Therefore, the importance of acquiring high quality data cannot be understated. The quality of the data is dependent upon their accuracy (*i.e.*, how close the measurement is to the actual concentration) and their reproducibility (*i.e.*, can the measurement be repeated within a certain precision, not only by the original researcher but also by all researchers using the same model). To this end, parameters that have been identified as sensitive in the proposed models (*e.g.*, MINTEQA2 with Model V and the metals translator approach) must be measured and collected using consistent, standardized methods, as well as methods capable of achieving sufficiently low detection limits while controlling contamination. To ensure consistency, a single method for each critical parameter must be required by the regulations. Consensus on such a method should be achieved through peer review before the regulations are released. One protocol that may be a good candidate, USEPA Method 1699 (USEPA 1995), was recently released. It contains recommendations for sampling and measuring metals at USEPA WQC levels and incorporates clean and ultraclean techniques.

In metal speciation, pH is a master variable. Careful quality control (Davison 1990; Durst *et al.* 1994) is required to obtain valid data. Where possible, pH values should be measured *in situ*, and multiple measurements should be obtained over diurnal periods to ascertain the variability.

Sampling variability
A potential advantage of modeling is that an assessment can be made of variability in chemical conditions at a given site brought about, *e.g.*, by variations in flow and pollutant discharges. For example, the effect of a ± 0.2 pH error on predicted free copper concentration was plus/minus a factor of two. In real natural waters, photosynthesis and respiration can cause much more profound variations in pH than pH measurement errors. The chemical speciation of copper(II) would be expected to be quite different at night or

early in the day, when pH values are lower, than it would be during the middle of the day, when pH is generally higher. This problem is especially severe in productive lakes. Chemical speciation models must therefore examine the effect of pH over its expected diurnal range.

Speciation and toxicity

Of what use are chemical speciation determinations or calculations in relation to toxicity? Allen and Hansen (1996) indicate that speciation can be used to decide which form of the metal would be bioavailable or toxic. If the concentration of the bioavailable form of the metal is modified by uptake, however, other metal species will react to reestablish equilibrium. Hence, for an equilibrium model, all metal species in equilibrium with the bio-reactive form of the metal can be considered bioavailable. Perhaps it is better to think of the bio-reactive form of the metal being correlated (not necessarily linearly) with toxicity. Thus speciation gives a guide to the potential of the metal to be toxic but does not indicate directly how much metal is available. As stated in the introduction, however, specific metal species concentrations may be buffered if master variables controlling the ligands and binding sites are more or less constant.

At the simplest level, the biological receptor is "just another ligand." Therefore, it is possible to formulate an equilibrium model of toxicity by including the gill receptor as a ligand, as proposed in Chapter 4. This is helpful because it allows competition by other components, *e.g.*, protons and calcium ions, to be included explicitly, thereby offering a means to explain their "protective" effects. The relevant speciation calculation would of course need to refer to the water in the vicinity of the receptor, which in the case of a fish gill would be buffered by excreted ammonia and carbon dioxide. However, one has to assume at least a local equilibrium if the concept is to be scientifically defensible.

Critique of the "Gill Model"

A proposal (Chapter 4) has been made to develop a "mechanistic" description of the toxicity of metal ions to fish. The approach is based on calculating the speciation of the metal ion in a solution that also contains a biological receptor site that is assumed to behave as another ligand. Because the proposed approach is essentially a chemical equilibrium model that would involve collection and interpretation of data from multicomponent chemical systems, it is appropriate to examine this approach from a chemical perspective.

The model is still under development, but in its current form (Playle *et al.* 1993a, 1993b; Janes and Playle 1995), it includes the following types of reactions, all of which are assumed to be at equilibrium:

Binding of the metal ion M (e.g., Cu^{2+}, Ag^+) with a specific receptor site X on the organism

$$M + X = MX \qquad K_{MX} = \frac{[MX]}{[M][X]}$$

At present, the receptor sites under investigation are those on a fish gill, but the approach could, in principle, be extended to include as yet undefined sites on invertebrates.

Competitive complexation of the metal ion in solution by protective ligands, which are described by reactions of the type

$$ML = ML \qquad\qquad K_{ML} = \frac{[ML]}{[M][L]}$$

The protective ligands could include DOC, simple inorganic ligands (OH^-, CO_3^{2-}, Cl^-), and simple organic ligands (oxalate, *etc.*).

Competitive binding of other cations at the receptor site, which inhibit binding of the metal ion M

$$Ca^{2+} + X^{n-} = CaX^{2-n} \qquad\qquad K_{CaX} = \frac{[CaX^{2-n}]}{[Ca^{2-}][X^{n-}]}$$

The competitive cations could include H^+ and major cations (*e.g.*, Na^+, Ca^{2+}).

To date, the most extensive datasets have been used primarily to describe binding of the metal to the gill. The binding has been carried out as a function of the concentration of metal ion M, competing cations, and protective ligands. In principle, the model could be extended to include the possibility for other toxic metal ions to bind at the receptor sites or other receptor sites. Most of the data in the recent literature on this subject (Playle *et al.* 1993a, 1993b; Janes and Playle 1995) are related to gill uptake of the metal ion only; the link to toxicity itself has not been explored extensively except by MacRae *et al.* (1997).

If it is successful, the potential advantages of this mechanistic approach are that
1) it can be used to aid in the development of new criteria documents that consider variations in chemistry,
2) it will provide criteria with a rational scientific basis, and
3) it will provide the link between measurable chemical parameters and biological effects.

The potential disadvantage of this approach is that the simple model may not in fact be sufficient to describe toxicity over a sufficient range of variations in solution composition. If this situation is not recognized, the model might be prematurely judged to be predictive, when in fact it is valid only over the range of solution composition used in model calibration. (Such a situation might arise, for example, if the protective effect of DOC was determined in one set of experiments, the protective effect of Ca^{2+} was determined in another set of experiments, but the effect of simultaneous variations of DOC and Ca^{2+} was not investigated.) On the other hand, if the model deficiencies are recognized and attempts are made to resolve these differences with *ad hoc* fixes, the model could lose its mechanistic basis, scientific credibility, and practical applicability. Certainly, precedents for both of these undesirable outcomes exist, as scientists have attempted simple mechanistic solutions to complex environmental problems.

While this mechanistic approach is certainly a valid long-term research topic, a simpler, more empirical approach as outlined in Chapter 4 (section "Understanding mechanisms of acute toxicity") is an alternative that warrants serious consideration for application now and in the long-term time frame.

Several aspects of the mechanistic approach require particular attention and critical review as research proceeds:

1) the neglect of kinetics (*i.e.*, the assumption of equilibrium in the solution and at the gill surface) in a system known not to be at equilibrium,

2) the link between the concentration of the target metal on the gill and the toxicity; and

3) experimental verification of the model for simultaneous variations in several water chemistry parameters. That is, the model should be internally consistent with all sets of data simultaneously.

The work to date could be characterized as "illustration of principle." The transition to "definitive proof," in which all of the three issues listed above are critically evaluated, should be made as soon as possible.

Dietary exposure to metals

Little is known about the forms in which metals are incorporated into components of aquatic food webs (Chapter 4, section "Understanding multiple exposure routes"). Some of the research on agricultural (*e.g.*, Chaney 1988) and human uptake studies (*e.g.*, Sandstead 1988) might give helpful information. However, fundamental research will be needed to investigate dietary exposure to metals, and an opportunity exists for fertile interchange of information and ideas between chemists and biologists.

Recommendations

The following recommendations are made by the Chemical Speciation and Toxicity Workgroup:

1) MINTEQA2 and MINEQL$^+$ would benefit from inclusion of Model V, and this modification would enable a wide range of users to carry out metal speciation calculations. Implementation is desirable in one year.

   - When merging Model V into other speciation systems, known deficiencies in the database (*e.g.*, MINTEQA2) need to be corrected.

   - Routine sensitivity analysis should be available for MINTEQA2 and MINEQL$^+$ incorporating Model V. Variability in data and model speciation constants would be the input for these estimations.

   - Biological labs using metal speciation in their studies could use WHAM.

2) WHAM and other shells incorporating Model V should be utilized to evaluate the potential for correlation of toxicity with specific aqueous metal species, given total dissolved metal concentration.

- Examination of existing datasets, containing total dissolved metal concentration and toxicity data, would indicate if adverse effects are discernible.
- Emphasis on instances where toxicity was/was not exhibited in the context of metal speciation would assist in evaluating regulatory policy.

3) A fundamental investigation is needed to relate pH and dissolved metal to suspended solids on the basis of measurable solid characteristics such as POM and oxyhydroxides. In the interim, to provide the link between dissolved metal concentration and total recoverable metal needed for transport and permitting purposes, we recommend the Kinerson (USEPA 1996) empirical K_d approach if the system is adequately characterized with respect to pH, TSS, hardness, salinity, and flow ranges.

4) Determination of dissolved metals is uncertain, and it is not clear that the filtration issue can be resolved. However, other techniques that measure dissolved species directly need to be assessed.
- In addition, methods used to acquire metals data should be standardized.
- Clean or ultraclean procedures should be used in all analyses.

5) WHAM and modified MINTEQA2 and MINEQL$^+$ models should be used comprehensively in the design of experiments and in the interpretation of data with respect to research on toxicology. Use of these models will permit easier integration of metal speciation and the proposed gill response model.

6) Collation of information and research on forms of metals in food web materials is needed for modeling metal accumulation and effects.

7) The term *bioavailability* is misused in the context of equilibrium speciation. The key point has to do with correlation of metal species to toxicity. Alternately, an explicit model may be used.

8) As work on the gill model progresses, the following points should be borne in mind to ensure a useful product:
- Extensive sets of data covering the ranges of concentrations of systems variables need to be developed to relate solution composition to toxicity.
- The model should be internally consistent simultaneously with all sets of data used in its development.
- Transition from the present "proof-of-principle" phase to a definitive model needs to be made expeditiously.

References

Allen HE, Hansen DJ. 1996. The importance of trace metal speciation to water quality criteria. *Water Environ Res* 68:42–54.

Allison JA, Perdue EM. 1994. Modeling metal-humic interactions with MINTEQA2. In: Senesi N, Miano TM, editors. Humic substances in the global environment and implications on human health. Amsterdam: Elsevier. p 927–942.

Buffle J, Leppard GG. 1995. Characterization of aquatic colloids and macromolecules: I. Structure and behavior of colloid material. *Environ Sci Technol* 29:2169–2175.

Chaney RL. 1988. Metal speciation and interactions among elements affecting transfer in agricultural and environmental food chains. In: Kramer JR, Allen HE, editors. Metal speciation: theory, analysis and application. Chelsea MI: Lewis. p 219–260.

Davison W, Zhang H. 1996. *In situ* speciation measurements of trace components in natural waters using thin-film gels. *Nature* 367:545–548.

Davison W. 1990. A practical guide to pH measurement in freshwaters. *Trends Anal Chem* 9:80–83.

Durst RA, Davison W, Koch WF. 1994. Recommendations for the electrometric determination of the pH of atmospheric wet deposition (acid rain). *Pure Appl Chem* 66:649–658.

Dzombak DA, Morel FMM. 1990. Surface complexation modeling: hydrous ferric oxide. New York: J Wiley.

Janes N, Playle RC. 1995. Modeling silver binding to gills of rainbow trout (*Oncorhynchus mykiss*). *Environ Toxicol Chem* 14(11):1847-1858.

Lee S-Z, Allen HE, Huang CP, Sparks DS, Sanders PF, Peijnenburg WJGM. 1996. Predicting soil-water partition coefficients for cadmium. *Environ Sci Technol* 30:3418–3424.

Luther III GW, Rickard DT, Theberge S, Olroyd A. 1996. Determination of metal (bi)sulfide stability constants of Mn^{2+}, Fe^{2+}, Co^{2+}, Ni^{2+}, Cu^{2+} and Zn^{2+} by voltammetric methods. *Environ Sci Technol* 30:671–679.

MacRae RK, Smith DE, Swoboda-Colberg N, Meyer JS, Bergman HL. 1997. Copper binding affinity of rainbow trout (*Oncorhynchus mykiss)* and brook trout (*Salvelinus fontinalis*) gills. *Environ Toxicol Chem* (in press).

Mouvet C, Bourg ACM. 1983. Speciation (including adsorbed species) of copper, lead, nickel and zinc in the Meuse River: observed results compared to values calculated with an equilibrium computer model. *Water Res* 17:641–649.

Playle RC, Dixon DG, Burnison K. 1993a. Copper and cadmium binding to fish gills: estimates of metal-gill stability constants and modelling of metal accumulation. *Can J Fish Aquat Sci* 50:2678–2687.

Playle RC, Dixon DG, Burnison K. 1993b. Copper and cadmium binding to fish gills: modification by dissolved organic carbon and synthetic ligands. *Can J Fish Aquat Sci* 50:2667–2677.

Sandstead HH. 1988. Interactions that influence bioavailability of essential metals to humans. In: Kramer JR, Allen HE, editors. Metal speciation: theory, analysis and application. Chelsea MI: Lewis. p 315–332.

Schecher WD, McAvoy DC. 1995. MINEQL[+]: a chemical equilibrium program for personal computers. Hallowell ME: Environmental Research Software. Version 3.01.

Serkiz SM, Allison JD, Perdue EM, Allen HE, Brown DS. 1996. Correcting errors in the thermodynamic database for the equilibrium speciation model MINTEQA2. *Water Res* 30:1930–1933.

Shuman MS. 1990. Carboxyl acidity of aquatic organic matter: possible systematic errors introduced by XAD extractions. In: Perdue EM, Gjessing ET, editors. Organic acids in aquatic ecosystems. Dahlem Konf. Wiley-Interscience. p 97–110.

Tercier HL, Buffle J. 1993. *In situ* voltammetric measurements in natural waters: future prospects and challenges. *Electroanalysis* 5:187–200.

Tipping E. 1993a. Modeling ion binding by humic acids. *Colloids Surf* 73:113–171.

Tipping E. 1993b. Modeling the competition between alkaline earth cations and trace metal species for binding by humic substances. *Environ Sci Technol* 27:520–529.

Tipping E. 1994. WHAM - a chemical equilibrium model and computer code for waters, sediments and soils incorporating a discrete site/electrostatic model of ion-binding by humic substances. *Computers & Geoscience* 6:973–1023.

Tipping E, Hurley MA. 1992. A unifying model of cation binding by humic substances. *Geochim Cosmochim Acta* 56:3627–3641.

Tipping E, Reddy MM, Hurley MA. 1990. Modeling electrostatic and heterogeneity effects on proton dissociation from humic substances. *Environ Sci Technol* 24:1700.

Tipping E, Woof C, Hurley MA. 1991. Humic substances in acid surface waters: modeling aluminum binding, contribution to ionic charge-balance and control of pH. *Water Res* 25:425–435.

[USEPA] U.S. Environmental Protection Agency. 1995. Whole effluent toxicity: guidelines establishing test procedures for the analysis of pollutants. *Federal Register* 60(199):53529–53544.

[USEPA] U.S. Environmental Protection Agency. 1996. The metals translator: guidance for calculating a total recoverable permit limit from a dissolved criterion. Kinerson RS, Mattice JS, Stine JF, editors. Washington DC: Office of Water. EPA 823-B-96-007. 60 p.

Environmental fate and transport

Jerald L. Schnoor, chair
John P. Connolly, Dominic M. Di Toro, Nico de Rooij, Miriam Diamond,
Russell S. Kinerson Jr., Donald B. Porcella, William L. Richardson, James F. Stine

Mathematical models of fate and transport of chemicals in the aquatic environment are the tools used to establish WLAs. Wasteload allocations for individual point source discharges or TMDLs for multiple sources are intended to ensure compliance with WQS, which in turn protect designated water uses and human and ecological health. Since the initiation of this process, the question has been "What modeling framework should be applied and to what level of physical, chemical, and biological resolution?" (Thomann and Mueller 1987).

Generally, permit writers do not have the time to assemble the requisite data to apply the most realistic and holistic models; as a consequence, simple dilution formulations of pollutant load divided by receiving water flow (W/Q) are used. This approach to developing permit limits has achieved significant improvements in the quality of U.S. waters. However, there are concerns that it 1) may not provide protection in the most cost-effective manner (*i.e.*, regulations are overprotective and result in costs without benefits) and/or 2) it may not adequately take into account the multiple uses and stresses placed on aquatic resources (Schnoor 1996).

In order to protect aquatic organisms that spend their life cycles in the water column and/or in the bed sediments, it is important to consider the chemistry of the metals and their transformation between free ion and complexed forms; to consider the partitioning of metals between water column, sediments, and fish; and to consider all the sources of metals loadings to the aquatic ecosystem (Figure 6-1). This involves greater modeling sophistication than is evidenced by the simple dilution equation (W/Q) that has commonly been used for setting permit limits. In order to consider all sources of metal, not just those discharged from a single permitted facility, a more comprehensive framework is needed. Fortunately, an appropriate framework, the watershed approach, has been suggested and is being adopted increasingly by states for water management, planning, and permitting of effluent discharges. Properly applied, the watershed approach accounts for all sources of the metal, point source discharges as well as nonpoint source discharges (including metals from municipal stormwater runoff), metals occurring naturally in soils, and atmospheric inputs of metals.

The goal is to improve present approaches for calculating effluent limits, taking into account the best science while maintaining the permit writer's ability to apply improved models accurately and efficiently.

State of the discipline in metals permitting

The workgroup developed criteria for evaluating models used for metals permitting and then evaluated existing models on the basis of those criteria.

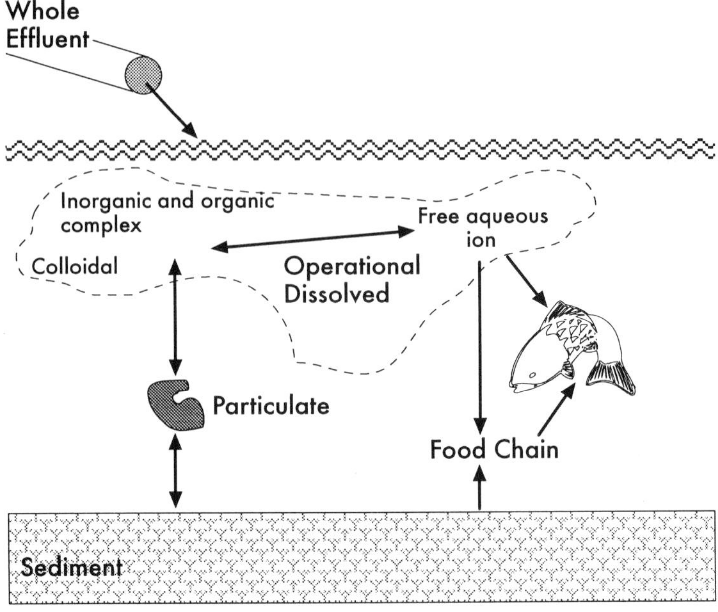

Figure 6-1   Schematic of fate processes for trace metals in an aquatic ecosystem

Evaluation of criteria for models in the permitting process

There are several criteria on which models used in permitting decisions should be evaluated:

- Appropriateness
- Accuracy
- Uncertainty
- Bias
- Feasibility
- State of scientific knowledge

The *appropriateness* of a model refers to the level of complexity that correctly addresses the problem but is not excessively complicated. The best model is the simplest one that adequately describes the situation. Simple models are necessary when dealing with the public and with decision-makers at all levels of government, and they minimize data required for applications. In the final analysis, models must be appropriate for the task and must result in an accurate representation of the problem.

Models should be accurate and simulate the actual exposure concentrations that one would observe in a given water body under known conditions. *Accuracy* refers to the "correct" concentration in space and time predicted by a model.

Bias refers to model predictions that are consistently too high or too low. This results in simulated exposure concentrations that are not indicative of the actual exposure to aquatic biota. Model results should be accurate and free from bias except in cases where data are sparse and a conservative approach is needed.

Model *uncertainty* is large where the state of the science is not well developed and when there are little field data to test a model. Repeated applications and testing of a model help to reduce uncertainty by increasing confidence that the model can handle a variety of situations.

Feasibility refers to the practical use of models in the permitting process. Costs, model difficulty, and time and data requirements should not be excessive for the task.

Lastly, the *state of scientific knowledge* represented by a model should be consistent with first principles and the literature regarding relevant fate processes. For example, if we know that a metal changes its speciation in the aquatic environment following its discharge and this in turn changes its toxicity, the model should be modified to account for the form of the metal that is likely to occur in the environment. Mathematical models are the only method that we have to estimate changes in water quality in space and time and to predict toxicity to aquatic biota.

Evaluation of existing models

Ambient concentrations of heavy metals in aquatic systems vary in space and time in response to fluctuations in mass loadings, the extent of dilution, and the rates of reactions that affect the transport and chemical form of the metal. Recognizing these fluctuations, various strategies have been employed to predict (model) the relationship between mass loading and ambient concentration.

Static dilution model. Static dilution (SD) is the most commonly used available model. Typically, a statistically defined low-flow condition is used with a maximum allowable concentration (*i.e.*, the water quality criterion concentration) to determine a maximum allowable mass loading. The concentration-loading relationship is computed at a single point that represents the location, downstream of the discharge, at which the metal is uniformly mixed over the cross-section (river or estuary) or volume (lake) of the water body. In almost all cases, the criterion concentration is calculated using a minimum hardness concentration for the water body (USEPA 1996). The main weakness of this approach is that it does not consider the frequency of occurrence of the condition that is modeled. The defined low-flow, the maximum mass loading, and the minimum hardness may occur simultaneously much less frequently than the once-in-three-year return period prescribed in the WQC. Other weaknesses include an inability to consider mixing zones, sediments, and cumulative effects.

Probabilistic dilution model. The probabilistic dilution (PD) model extends the SD model to the full range of dilution and mass loading conditions. It eliminates the uncertainty that exists with the SD model regarding the meaning of the loading limit because it calculates the allowable mass loading consistent with a WQS that has a defined, allowable frequency of exceedance (Reckhow and Chapra 1983). Additionally, it achieves this reduction in uncertainty with few additional data needs. The statistics of mass loading and of the flow are requirements of both models. The mass loading statistics are needed to translate a maximum allowable loading to an average loading. The statistics of flow are needed to define the low-flow condition. The additional data requirements of the PD model pertain to correlation among the variables in the dilution equation (Schnoor 1996).

The main weakness of the PD model is in its application to the multiple discharge problem. Because it is a point model, it does not account for changes in concentration that occur between discharge locations due to the various fate processes. Thus, it tends to compute conservative (high) estimates of concentration, with bias increasing with the spatial scale of the problem and the number of discharges considered.

As with the SD model, the PD model cannot address the issue of sediment contamination.

Spatially resolved steady-state model. Spatially resolved steady-state (SRSS) models are used in much the same way as the SD model: to determine maximum allowable loadings at the statistically defined low-flow condition. These models typically consider sorption so that the loss of metal due to settling to the sediment may be included (Thomann and Mueller 1987). These models can include chemical equilibrium calculations to describe speciation and precipitation/dissolution as described by Schnoor (1996). (Note that dilution models can also incorporate sorption and chemical equilibrium principles, although in practice they commonly do not.) Spatially resolved steady-state models provide advantages in that they account for the interaction among discharges, and they provide a spatial profile of concentration that can be used to examine mixing zone issues and the spatial extent of any standards violation. They suffer the same major weakness of the SD model: they do not provide information about the frequency of occurrence of the modeled condition.

An additional potential weakness of the SRSS models relates to their capabilities to simulate sediment-water interactions. This weakness is the result of the differing time scales associated with metals in these media. The water column responds rapidly to changing inputs; the sediment responds much more slowly. Consequently, metal in the sediment is not necessarily at steady-state with any particular condition in the water column. If a steady-state condition exists, the sediment is at steady-state with the long-term average condition in the water. This condition is probably poorly described by the low-flow water column condition.

Spatially resolved time-variable model. The spatially resolved time-variable (SRTV) model has been applied only to the more complex problems. When sediments are considered, these models typically require a large computational effort and detailed information about the dynamics of water and sediment transport. The requirements are smaller if the model is restricted to the water column, although the capability of the model to simulate the metal under high-flow conditions may be compromised by the absence of sediment.

The main advantages of the SRTV model are that 1) it can correctly describe sediment-water interactions for the net effect of multiple discharges and 2) it can describe the time response of the water body to initial conditions and changes in the mass loading of metals. Thus, this type of model is useful for estimating the relationship between metals discharge and SQC and for determining the benefits achieved by reductions in metals loading.

Existing SRTV models include empirical descriptions of equilibrium metal sorption to particulate matter and DOM. Simple descriptions of the equilibrium interaction of the metals with AVS have been included. In addition, recently developed sediment sub-models that describe the kinetics of metal-sulfide interactions could be readily incorporated in the SRTV models.

Recommendations

This section presents the workgroup's recommendations for improvements that can easily be brought to the permitting process and some research needs that will result in future improvements.

Near-term recommendations

1) Adopt a probabilistic modeling approach for riverine systems rather than applying SD methods.
2) Design models to protect both the water column and sediment.
3) Issue permits on a watershed basis (where applicable).
4) Continue the use of translator methodology (USEPA 1996) for permits that use the operationally defined, dissolved fraction of metal as indicative of the toxic species.

Long-term recommendations

1) Adopt a watershed approach using a GIS to manage multiple discharges and nonpoint source loadings (near and far field).
2) Include time dependency in spatially resolved models to predict sediment quality, diurnal variations, storm events, and resuspension.
3) Account for multiple routes of chemical exposure to biota, *e.g.*, inclusion of a food chain model.
4) Test models under a variety of field conditions (long- and short-term, near and far field).

Research and development needs

1) Develop an NEDSS to support modeling efforts toward establishing discharge permits that use more comprehensive and scientifically credible approaches, including the watershed approach.

2) Initiate NMPS to gather data for site-specific validation of transport and fate models used in permitting.

Implementation

The implementation of modeling to the permitting process can be viewed as a matrix with various aquatic ecosystems across the top (columns) and different modeling approaches arranged down the side (rows), from the least to the most complicated (Table 6-1). We recommend that the PD model be the starting point for modeling riverine environments (streams with one reach and multiple reaches). In the near term, the probabilistic approach in watersheds could be applied to cases where discharges from all effluents and nonpoint sources are considered in the permitting process. The probabilistic approach is less applicable in lakes, estuaries, harbors, or for coastal discharges, because the systems are not advection dominated, and because it is difficult to establish probability density functions for upstream flow and concentration in these cases. For such systems, it is best to use an SRSS model or an SRTV model. The SRTV approach is powerful either when sediment dynamics must be included in the calculation of water quality concentrations or for the purpose of examining SQC. An SRTV approach could be taken even further, including kinetics of adsorption/desorption and redox processes and chemical speciation, but this would pertain to research models and not to the permit process.

Near-term implementation. The tools for using probabilistic modeling to calculate water quality permit limitations are sufficiently developed for routine use. Probabilistic modeling has the potential for providing less biased permit limits than does using simple SD models. Key variables such as low flow, maximum mass loading, and minimum hardness may occur simultaneously at a different frequency than the once-in-three-year period prescribed by the WQC.

Discharge limitations should be designed to protect both the water column and the sediments. The existing regulatory approach is to compare the discharge limitation based on the WQC with the limitation based on the sediment criterion and to apply the stricter limit. An unresolved question is whether meeting the WQC will also protect the sediments. This key question should be resolved with further research.

Issuing permits on a watershed basis is encouraged. Existing requirements to develop TMDLs and allocate watershed load-bearing capacity among point and nonpoint sources should be implemented. Models are available to facilitate this activity. Watershed permitting should consider loads from point and nonpoint sources (including atmospheric deposition) and sediment/water exchange when feasible.

Bioavailability should be accounted for by partitioning between particulate and dissolved forms and by providing the translators (USEPA 1996) needed to develop the total recoverable permit limit.

Future implementation. The following recommendations are intended to take us from what we can accomplish in the short term to achieving our long-term objectives.

Table 6-1   Approaches to modeling metals in a variety of aquatic ecosystems

Approach[a]	Stream reach/ one discharge	Multiple reach and discharge	Water-shed (PS & NPS)	Whole lakes/ pelagic zones	Littoral zones of lakes	Estuaries and harbors	Coastal marine
1. SD[b] (W/Q)							
2. PD[b]	Recom-mended	Recom-mended	Recom-mended				
3. SRSS[b]				Recom-mended	Recom-mended	Recom-mended	Recom-mended
4. SRTV[c]	Recom-mended	Recom-mended	Recom-mended	Recom-mended	Recom-mended	Recom-mended	Recom-mended
5. SRTV kinetic/ speciation	Possible-future probable	Possible-future probable	Possible-future probable	Possible-future probable	Possible-future probable	Possible-future probable	Possible-future probable

SD = static dilution; PD = probabilistic dilution; SRSS = spatially resolved steady-state; SRTV = spatially resolved time-variable
[a]Add processes as needed
[b]Water column only
[c]Sediments included

The workgroup recommends that NPDES permits be issued within a watershed perspective. This more holistic approach would replace the present piecemeal approach, in which models are used to determine WLAs separate from other stresses. To achieve the goal of developing more complex models that can be easily used by the permit writer, we must maintain user-friendly models that are also scientifically defensible.

Developing a watershed approach builds on existing technologies and experience. Our recommendations are as follows:

1) Spatial consideration: Link the watershed through GIS so that geographically specific load allocations and estimates of nonpoint source loadings are calculated and managed. The watershed model should simulate sediments as appropriate. It is important to appreciate that metals will tend to accumulate in sediments at downstream locations that are geographically removed from emission sources.

2) Temporal consideration: Include temporal resolution as a time-dependent model. Time dependency is important for capturing instances when water quality may be compromised due to low-flow conditions.

3) Include a food chain component (Thomann and Mueller 1987) to account for all routes of chemical exposure to organisms. This will provide the means for accounting for chronic toxicity and nonlethal effects resulting from chemical transfer through food chains.

4) Include a process whereby models are subjected to medium- and long-term scrutiny (*i.e.*, accuracy, precision, and bias). This should include collection of field data for comparison to model predictions and peer review.

Research and development

The research community must proceed beyond the immediate and short-term goals related to permit writing and anticipate future requirements of the long-term regulatory needs. Regulators should be sensitive to the longer range vision by encouraging and supporting this more basic research that leads to a more holistic and broader scale of environmental problem solving. In this regard, there are two recommendations. First, models should be made general to most chemicals, volatile and nonvolatile (inorganics and organics), and should include those which transform to different redox states and between organic and inorganic forms. Assessment of eutrophication and acidification issues will require inclusion of nutrients (*e.g.*, phosphorus, nitrogen) and acidic species. Second, multimedia models are necessary to treat a diversity of chemicals and loading sources.

Research focuses on improving the scientific credibility and reliability of models capable of simulating and predicting concentrations of metals in aquatic systems, while development focuses on efforts to transfer research results and findings to a problem-solving and/or regulatory setting. The process of conducting research for the purpose of improving environmental regulation is the key. Coordinated, directed, and site-specific research projects are needed to assure realism. A system for organizing data, models, and information assures an efficient application.

Development recommendations

The following recommendations are related to development issues:

1) Develop and implement an NEDSS. To facilitate the use of more sophisticated, more scientifically credible models in the permitting process, an NEDSS needs to be developed and implemented. This system would combine existing technologies such as GIS, environmental databases, and computer modeling programs along with bibliographical information and fate-and-effects databases to generate input datasets for model operation. Such a system would link the regulator, the regulated, and the scientist in an interactive and ongoing process using the Internet to post information, apply models and databases, and update those involved. The system must be user friendly and must make the regulatory process efficient and reasonable. This NEDSS needs to be developed and supported on a continuing basis, most likely by an applied research laboratory or institute or perhaps by a team of several laboratories.

2) Implement NMPS. Prototype modeling studies conducted in a range of geographic locations should calibrate and test the validity of new modeling frameworks. Investigators representing a range of disciplines (chemists, toxicologists, and engineers) would conduct coordinated research and monitoring and would produce comprehensive datasets for a range of environmental parameters over at least an annual hydrological cycle. These studies would be

conducted at a number of sites representing the range of environmental systems: rivers, watersheds, harbors, lakes, and coastal areas. Once tested and peer reviewed, prototype models would be incorporated into the NEDSS.

Modeling research priorities

1) NMPS will develop algorithms for models and test the validity of watershed models.

2) Hypotheses will be tested that link metal exposure to effects at the target organ, the results of which would be incorporated into model algorithms.

3) Chemical speciation will be incorporated in the watershed model framework.

4) Models will be developed that include sorption/desorption kinetics and speciation. Currently, metals models use equilibrium sorption assumptions to compute dissolved and particulate fractions. Previous studies have shown that these assumptions produce erroneous results because desorption from either point sources or from resuspended sediment is not instantaneous.

5) Fate, transport, and interactions of multiple chemicals in whole effluents will be included in the model framework.

Process and experimental research priorities (in support of model development)

Sediment-water/water-particle exchange:

1) The relationship between sediment criteria and WQC will be investigated in future research. The basic question is, "Will meeting WQC concurrently satisfy SQC?" Also, both process and modeling research is needed to determine the relative importance of "in-place" contamination versus new contamination from current discharges.

2) Understanding the mechanisms and rates of metals release from resuspended sediment is needed (Di Toro 1997).

3) A more thorough understanding of resuspension rates and amount of material resuspended under high-shear stresses during major storm events is required. Consideration of bioturbation and properties of mixtures of fine-grained and coarse-grained sediments should be included. Implicit is the developmental need for methods to accurately map and characterize sediments and other characteristics of watersheds using GIS.

4) A more thorough understanding is required for contaminant partitioning to aquatic particulate matter.

5) *In situ* modeling and research is needed to better understand dynamics at the sediment/water interface. In some large aquatic systems, there is a dense, suspended solids layer near the sediment interface referred to as the "nephloid layer." Little information is available regarding its origin or its relationship to contaminant fate and transport.

Nonpoint sources:

1) Atmospheric deposition must be quantified to determine its importance as a source of trace metals to watersheds, particularly in urban areas.

2) Urban runoff and combined sewer overflows need to be characterized better for metals discharges.

3) Agricultural runoff, including nutrients and suspended material, is changing as a result of farm practices. We need a better understanding of metals emission factors as a result of land use changes and farm practices.

References

Di Toro DM. 1997. Sediment flux modeling. New York: J Wiley. (in press).

Reckhow KH, Chapra SC. 1983. Engineering approaches for lake management. Volume 1. Boston MA: Butterworth. 340 p.

Schnoor JL. 1996. Environmental modeling. New York: J Wiley. 682 p.

Thomann RV, Mueller JA. 1987. Principles of surface water quality modeling and control. New York: Harper & Row. 644 p.

[USEPA] U.S. Environmental Protection Agency. 1996. The metals translator: guidance for calculating a total recoverable permit limit from a dissolved criterion. Kinerson RS, Mattice JS, Stine JF, editors. Washington DC: Office of Water. EPA 823-B-96-007. 60 p.

Appendix: Discussion-Initiation Papers

Development and use of aquatic toxicity criteria for heavy metals

Alexander Viteri, Jr.

"One of the primary missions of the Society of Environmental Toxicology and Chemistry (SETAC) is to promote the use of multidisciplinary approaches to examine the impacts of chemistry and technologies on the environment. An equally important strength is the Society's commitment to balance the interests of academia, business, and government and to facilitate collaboration across the sectors. Such a collaborative effort surely must lead to a progressive and enlightened approach to addressing issues of concern to environmental toxicology and chemistry" (Benson 1995).

This presentation includes a brief history of the development and use of aquatic toxicity criteria for heavy metals. In keeping with the above mission, SETAC has challenged this Workshop to advance a more enlightened and progressive approach towards development and use of heavy metals criteria and to identify research priorities needed to accomplish this. The 1972 Federal Water Pollution Control (FWPC) Act states that "It is the national goal that whenever attainable, an interim goal of water quality which provides for the protection and propagation of fish, shellfish, and wildlife, and provides for recreation in and on the water be achieved by July 1, 1983" (FWPC 1972a). The Act devised simple and direct approaches to get this done, including a mandate "that EPA publish water quality criteria accurately reflecting the latest scientific knowledge on the kind and extent of all identifiable effects on health and welfare, which may be expected from the presence of pollutants in any body of water." Before passage of the Act, states adopted WQC that reflected their level of interest in water pollution, how well toxicity data matched their interests, and the impact that resulting criteria could have on the state's economy. Because of this varied approach, it was common for a river to have many differing WQC to protect the same recognized use. Although not a major economic factor, a state's water criteria helped to influence the selection of new factory locations. The advantage of funding USEPA to develop nationally based WQC was that it established a uniform means for developing improved criteria for protecting water uses and placed that responsibility in the hands of an impartial federal agency, neutralizing any economic advantage enjoyed by states using more relaxed criteria.

The Act also created NPDES permits. Originally the FWPC Act required all discharges into navigable waters to get one of these permits. The Act imposed progressively increasing discharge limits uniformly across any given industry (FWPC 1972b). The USEPA developed and imposed effluent limits for each industry based on results of national surveys that identified the capability of an industry to achieve respective treatment levels. It then matched that information with the economic stability of the industry. Using this approach, most industries reduced the amount of pollution being discharged by between 90% and 97%. The last of these effluent limits addressed metal toxicity for the first time. Otherwise, effluent limits did not attempt to reconcile themselves with pollution levels needed to protect water uses. So, having accomplished the first major hurdle of the FWPC Act, the USEPA began replacing effluent limits with water quality-based discharge limits in NPDES permits.

Despite this and other major achievements, the USEPA lagged behind in developing criteria for toxic substances. The National Resource Defense Council (NRDC) sued the USEPA to begin establishing toxic criteria for a list of identified priority pollutants. The NRDC won the suit (*NRDC v. Train* 1976). The court required the USEPA to establish criteria for 129 priority pollutants within a specified time. It was during this period that the USEPA developed and utilized its approach for establishing specific complex pollutant water criteria. These criteria were based on the most protective approach found. Here is one example: "Copper: For freshwater and marinewater aquatic life, 0.1 times a 96-h LC50 as determined through non-aerated bioassay using a sensitive aquatic resident species. A summary of acute toxicity data is given in Table 5. In checking this table, the reader should consider the species tested, pH, alkalinity, and hardness if alkalinity is not given (in most natural waters, alkalinity parallels hardness). In general, salmonids are very sensitive and the centrarchids are less sensitive to copper" (USEPA 1976, p 54). Most of the toxicity tests listed extracted metals from particulate sediment and then ran bioassay tests using a mixture of the extracted fraction and purified lab water; toxicity was then reported as total recoverable heavy metals. This provided a conservative approach for protecting against fate processes that could increase the toxicity of a heavy metal downstream.

The approach used for heavy metals was very much parallel to the approach that the USEPA used to develop criteria for other similar complex elements, such as chlorine. "Chlorine in the free available form reacts readily, combing with nitrogenous organic materials to form chloroamines. These compounds are toxic to fish. Chloroamines have been shown to be slightly less toxic to fish than free chlorine, but their toxicity is considered to be close enough to free chlorine that differentiation is not warranted" (FWPC 1972b).

With development of toxicity criteria, NPDES permits matured from using effluent limits to using pollutant-specific water quality-based limits. National Pollutant Discharge Elimination System permits are now moving away from specific water quality-based criteria towards WET. All the while, the science of environmental toxicology and chemistry has had a difficult time understanding how development of heavy metal criteria progresses through these transitions. In the early 1990s, the USEPA (1995) issued interim guidance on heavy metals criteria suggesting that a state could use either the dissolved fraction or total recoverable method for establishing water criteria limits. Immediately after this, the USEPA's National Toxic Rule required the use of total recoverable metals in NPDES permits stating that using an approach similar to the AVS was best, but there were articulated problems with the AVS method's ability to work well with copper-contaminated sediments. In addition, the USEPA stated that the AVS method/theory does not work well for metals that do not generally associate with sulfides in sediment such as chromium (Chapman 1995). About the same time, USEPA Headquarter's Office of Science and Technology stated that states should use the dissolved fraction method to establish water criteria (Prothro 1993).

Several associations and states sued the USEPA over the resulting confusion—USEPA lost. The court required that the USEPA issue an administrative stay of aquatic life WQC contained in 40 CFR 131.36(b) for the following metals: arsenic, cadmium, lead, mercury (acute only), nickel, selenium (salt water only), silver, and zinc. Simultaneously, the USEPA identified its intent to include criteria for metals based on a total recoverable guidance value multiplied by a conversion factor, to convert the total recoverable value to the dissolved form of the metal specified in the Office of Water's metals policy (USEPA 1995). In order to do this, the USEPA created conversion and translation factors. So, if a state's WQS are based on the dissolved fraction of heavy metals, USEPA requires states to use conversion factors to change total recoverable metal values being proposed in a draft NPDES permit into equivalent dissolved fraction levels and then use transition factors to convert dissolved fraction values back to total recoverable. For example, states that use dissolved fraction to establish heavy metal water criteria must use a mathematical factor to convert the suggested total recoverable discharge level into an equivalent dissolved fraction value. This is done so the states understand the effects the proposed discharges may have on receiving water uses. Because of partitioning that occurs when effluent enters a receiving water body, a translation value is used to convert resulting receiving water dissolved fraction values back to total recoverable effluent levels.

"Sola Dosis Facit Venuenum" Paracelsus, 1493–1541: There is no material which is not a poison depending on the dose (Deichmann *et al.* 1986).

The viewpoint of most states is that the USEPA's approach may be a policy-based one that reflects limited use of chemistry or mechanistic toxicity. The current approach creates an exacerbating permitting process. The approach makes scientific explanation of metals criteria and related proposed permit discharge limits very difficult. Furthermore, I am suspicious that the difference between measured total recoverable metals and dissolved fraction metals in any given sample is meaningless, given the recent shift toward use of whole effluent toxicity.

Some recommendations for the workshop to consider:

- Review the difference in toxicity between total recoverable and dissolved fraction as they have been reported traditionally. Compare this difference to the ability of field studies to distinguish a corresponding difference in toxicity on immediate or downstream water uses.

- Given the above, consider if the difference in toxicity between total recoverable and dissolved is meaningful.

- The accuracy of analytical methods used to detect heavy metals in marine and brackish waters is questionable. Consider if the USEPA should be asked to determine the difference between using total recoverable and dissolved fraction in marine or brackish waters. Consider if the difference in total recoverable and dissolved fraction values in marine or brackish waters is within the range of analytical accuracy for saline waters.

- Consider recommending that the USEPA field-verify laboratory toxicity results before using lab results to develop WQC.

References

Benson WH. 1995. Better science makes for better decisions. *Environ Toxicol Chem* 14(11): 1811–1812.

Chapman PM, Thorton I, Persoone G, Janssen C, Godtfredsen K, Z'Graggen MN. 1996. International harmonization related to persistence and bioavailability. *Human Ecol Risk Assess* 2: 393-404.

Diechmann WB, Henschler D, Holmstedt B, Keil G. 1986. What is there that is not poison? A study of the *Third Defense* by Paracelsus. *Arch Toxicol* 58:207–213.

[FWPC] Federal Water Pollution Control Act. 1972a. P.L. 92-500. 92nd Congress, S.2770, October 18, 1972. Also titled Clean Water Act.

[FWPC] Federal Water Pollution Control Act. 1972b. Section 301 of P.L. 92-500. 92nd Congress, S.2770, October 18, 1972. Also titled Clean Water Act.

[NRDC] National Resource Defense Council v. Train 8 E.R.C. (D.D.C. 1976). Modified 12 E.R.C. 1833 (D.D.C. 1979).

Prothro MG. 1993 Oct 01. Office of Water policy and technical guidance on interpretation and implementation of aquatic metals criteria. Memorandum from Acting Assistant Administrator for Water to Water Management Division Directors. Washington DC: USEPA Office of Water. 7 p.

[USEPA] U.S. Environmental Protection Agency. 1976. Quality criteria for water. Washington DC: USEPA Office of Water and Hazardous Material.

[USEPA] U.S. Environmental Protection Agency. 1995 Apr. EPA issues the final water quality guidance for the Great Lakes system. Water Quality Criteria and Standards Newsletter. Washington DC: USEPA Office of Water. EPA/823/N-95/001.

North Carolina's alternative approach to implementation of standards for toxic metals

Larry W. Ausley, Dianne M. Reid

The national process of implementation of WQC and standards for metals has been wrought with controversy, mostly over the appropriate measure of and protection against environmental toxicity of the widely varying chemical forms in which metals occur in wastewater and surface waters. Regulators have attempted to base standards on several analytical techniques using a variety of assumptions about the bioavailability and fate of metals. This paper presents an overview of regulatory history of WQS for metals and an alternative approach to metals standards that has been implemented by the State of North Carolina. North Carolina's alternative approach is viewed by many to represent an effective and efficient method of addressing many of the concerns that, to date, still hamper widespread adoption of numerical WQC for metals.

History

Implementation of metals standards as "total metals" and "total recoverable metals" provided levels of protection against known effects of metals on aquatic species because, presumably, all metal present was assumed to be in its most toxic, bioavailable form, whether this was the case or not. These standards, also presumably, were effective in dealing with potential fate of metals through the processes of organic and inorganic ligature and resuspension to and from sediments. For the more ubiquitous metals like copper and zinc, though, the total and total recoverable metals standards were and are very difficult to achieve for many NPDES dischargers.

In the early 1980s, discussion began on whether realistic prediction of environmental effect could be better achieved by consideration of only the bioavailable, toxic fraction of the concentration of total metals. Various analytical chemical techniques were suggested to quantify various chemical forms of metals and their compounds. These techniques include total and total recoverable, acid soluble, and dissolved (USEPA 1992). The total, total recoverable, and acid soluble methods all use sample acidification and various temperature increases for totalizing metals concentrations that are or could become dissolved in an aquatic environment. Dissolved metals measurement and formulae for translating total recoverable measurements to dissolved concentrations are intended to estimate the assumed bioavailable fraction of metals in a sample. Chemical measurement techniques meant to estimate bioavailability are severely limited by the very nature of the assumptions that must be applied, for example, the assumption that the metal species measured are in forms that are toxic to aquatic species at the point of collection.

The regulatory trend over the past several years has been toward estimation of concentrations of immediately bioavailable forms with the use of dissolved measurement and dissolved translation approaches. Water-effects ratios take these approaches another step by establishing a discharge-specific translation of a numeric criterion based on observed toxicity of an effluent/receiving water matrix.

All of these current approaches still retain a disconnect in regulatory logic by maintaining the assumption that a surrogate measure of toxicity is necessary or even desirable. It seems much more logical to test directly the assumption of toxicity with reliable, cost-efficient, and now promulgated (USEPA 1995) aquatic toxicity methods (USEPA 1993, 1994a, 1994b), which make those direct measurements of bioavailability. The national paradigm of WQS has historically been one of chemical-specific criteria due, in part, to the lack of nationally recognized biological monitoring tools that could be broadly implemented in a regulatory context. The formation of the national policy of independent application of chemical and biological criteria was also driven to some degree by the uncertainty created by the lack of biological methods. This was a period when few regulatory agencies had implemented the biological tools necessary to broadly assess the effects of toxicants. Those situations have changed. We must now reevaluate the efficacy of the tools we use and policies that restrain use of those which prove to be most effective. Current criteria for several metals are based exclusively on results of aquatic toxicity analyses—tests that are similar or identical to those that *could* be used directly to remove assumptions about bioavailability in their matrix.

When criteria were based on total recoverable metals, the argument could be made that they provided conservative protection against fate processes that could increase bioavailability of metals temporally or spatially from the point of discharge. Current policy (the recommendations to use dissolved metals criteria, translators, or WERs) changes that viewpoint in providing protection at the point of discharge. The policy does so without the ability to make a very important consideration—what is the effect of the addition of multiple metals and other constituents, even below criterion levels?

North Carolina's approach to standards

In recognition of matrix effects the aquatic environment can have on the bioavailability and toxicity of some elements and ions, the North Carolina Environmental Management Commission adopted unusual, if not unique, state WQS for the metals copper, zinc, silver, and iron, as well as for chlorine and chloride. These "Action Level" WQS are defined in N.C. Administrative Code (NCDEHNR 1995) to include those substances "...which are generally not bioaccumulative and have variable toxicity to aquatic life because of chemical form, solubility, stream characteristics, or associated waste characteristics." These standards, listed in Table 1, do indeed create numerical criteria for these substances. In contrast to nationally recommended criteria, though, the standard is not established as mass per volume but rather implicitly as toxic units, *i.e.*, the standard is one chronic toxic unit.

In essence, these standards establish that water quality is not impaired if the concentration of an action level substance in an effluent is not toxic under critical dilution conditions (7Q10, permitted maximum flow). If an effluent is toxic and the action level parameter is identified as a causative toxicant, the permit may be reopened and limits established to meet the action level (for metals, as total recoverable concentration).

Table 1 North Carolina water quality action level standards

Parameter	Freshwater aquatic life standard	Tidal saltwater aquatic life standard
Chloride (mg/l)	230	N/A
Chloride, total residual (µg/l)	17	N/A
Copper (µg/l)*	7	3
Iron (µg/l)	1000	N/A
Silver (µg/l)	0.06	0.1
Zinc (µg/l)	50	86

*All metals measured as total recoverable

Though the USEPA had conditionally accepted North Carolina's use of these standards, the agency subsequently disapproved the implementation of copper and zinc in 1993 on the basis that they did not adhere to the policy of independent applicability of standards. The state and the USEPA are still in negotiation over the standards.

In 1987, North Carolina adopted the policy that all major and complex minor NPDES dischargers would receive WET limitations, assuming that this set of facilities, at least, had a reasonable potential to violate WQS for toxic substances. As such, this group of facilities would contain the same subset that would otherwise have had specific metals limitations. Along with WET limits and other chemical-specific limitations intended to protect resources in other ways, additional biological monitoring programs including macroinvertebrate population surveys, fish population surveys, phytoplankton and macrophyte population surveys, and fish and shellfish tissue monitoring round out a balanced North Carolina regulatory program which routinely utilizes the strengths of each tool to assure protection.

Benefits, drawbacks, and limitations

The significant benefit of North Carolina's implementation of action level WQS is the effectiveness by which it meets the intent of the criteria, *i.e.*, protecting against the discharge of toxic levels of toxic substances. By broadly applying WET limitations in these permits, the intent of specific metals criteria is met, but more so, the reality of combined chemical effects and matrix effects is addressed, and the need for extensive specific chemical monitoring is minimized.

Aquatic toxicity analyses provide some additional benefits of water quality protection not realized by chemical-specific methods. One of the most important of these benefits is measurement of combined effects of all toxicants in the solution, whether those effects are additive or subtractive, even at combined levels below those considered safe as individual criteria (Spehar and Fiandt 1986; Kraak *et al.* 1994). A transition away from the more conservative measure of total recoverable metals has also been made more feasible by the publication of sediment toxicity and bioaccumulation methods (USEPA 1994c). Use of these direct measures of bioavailability in the sediments would remove yet another layer of possible over-conservatism.

Whole effluent toxicity approaches to application of WQS through NPDES permitting retain most of the protective aspects of a chemical-specific approach. They still provide predictive protection to some low-flow condition, *e.g.*, 7Q10. With standards now being considered based on dissolved and/or bioavailable fractions of metals concentrations, the fraction measured is equivalent. If toxicity measurements use species as sensitive as species on which criteria were based, then protection is equivalent. If toxicity is measured using dilution water required by USEPA toxicity protocols (20% dilute mineral water or moderately hard reconstituted water), then protection against metals toxicity will likely be conservative due to the lack of dissolved organic substances, humics, and suspended solids, which would likely decrease environmental toxicity of those metals (Sprague 1985). In this aspect, North Carolina differs from USEPA protocols in allowing surface water to be used as a diluent in WET tests yet requiring lower total hardness, simulating local (state) surfacewater conditions. USEPA's current guidance seems to validate such site-specific consideration.

Dissolved criteria, total recoverable criteria with a dissolved translator, WERs, and WET limitations are all predicated on toxicity of a metal at the time of measurement and do little to address issues of environmental fate of non-bioavailable metal compounds. Conservative criteria based on total metal or assessment of compounds with potential to become toxic (*e.g.*, acid soluble) are needed to address environmental fate. These predictions will still be based on effects of a single metal in isolation of its environment. Alternatively, direct monitoring of protective assumptions has become feasible through standardized sediment toxicity methods which can, again, assess the combined effects of all constituents of sediment samples.

A significant benefit of biological analysis is the analytical accuracy of measured effect that removes a multitude of assumptions inherent in specific chemical criteria. As metals criteria and permit limits approach, or even go below, the analytical detection levels of current techniques and technology, analytical precision also becomes a significant factor in implementation. As concentrations approach detection level, precision is expected to decrease (USEPA 1991). The precision of chronic aquatic toxicity analyses is estimated at their detection levels. Thus, the precision of these tests, which are currently considered equivalent to those of chemical parameters derived well above detection levels, is better at metal concentrations near their chemical-specific lower detectable concentrations.

For metals like mercury, where significant bioaccumulation may be more problematic than immediately observable toxic effects, short-term WET analyses are not appropriate as protective tools. In these cases, the predictive protection provided by chemical-specific criteria and limitations, which are based on different assumptions, is more appropriately applied. For that reason, metals such as mercury are not considered by North Carolina for action level WQS.

Summary

Environmental regulators are rightly being asked to provide reasonable and realistic protection for aquatic resources without placing undue burdens on the regulated community. Many currently utilized chemical-specific criteria, including those for some metals, fall short of providing those realistic levels of protection due to many potentially overprotective assumptions inherent in their use. The USEPA (1992) has stated that "because of the complexity of metal chemistry, there is no one chemical analytical method that can accurately determine the metals that are bioavailable and toxic." In the past, the use of chemical-specific criteria was driven by lack of better tools by which to achieve the protection of aquatic resources. We now have the biological tools to remove many of the unnecessary assumptions about both immediate and far-afield effects of complex waste water, such as sediment accumulation, bioaccumulation, and sublethal effects. These tools include standardized WET analyses, sediment toxicity and bioaccumulation analyses, rapid bioassessment protocols, and fish and shellfish tissue analyses. We must begin to use those tools to provide realistic, effective, and efficient protection of our resources.

References

Kraak MHS, Lavy D, Schoon H, Toussaint M, Peeters WIIM, van Straalen NM. 1994. Ecotoxicity of mixtures of metals to the zebra mussel *Dreissena polymorpha*. *Environ Toxicol Chem* 13:109–114.

NCDEHNR. 1995. Administrative Code Section § 15A NCAC 2B.0211(b)(4). NCAC 2B.0200-Classifications and water quality standards applicable to surface waters of North Carolina. Raleigh: North Carolina Environmental Management Commission.

Spehar RL, Fiandt JT. 1986. Acute and chronic effects of water quality criteria based metal mixtures on three aquatic species. Duluth MN: USEPA Environment Research Laboratory. EPA/600/S3-85/074.

Sprague JB. 1985. Factors that modify toxicity. In: Rand GM, Petrocelli SR, editors. Fundamentals of aquatic toxicology. Hemisphere Publishing. p 124-163.

[USEPA] U.S. Environmental Protection Agency. 1991. Technical support document for water quality based toxics control. Washington DC: USEPA Office of Water. EPA/505/2-90-001.

[USEPA] U.S. Environmental Protection Agency. 1992. Interim guidance on interpretation and implementation of aquatic life criteria and metals. Washington DC: Office of Science and Technology, Health and Ecological Criteria Division.

[USEPA] U.S. Environmental Protection Agency. 1993. Methods for measuring the acute toxicity of effluents and receiving waters to freshwater and marine organisms. 4th ed. Cincinnati: USEPA Office of Research and Development. EPA/600/4-90/027F.

[USEPA] U.S. Environmental Protection Agency. 1994a. Short-term methods to estimate the chronic toxicity of effluents and receiving waters to freshwater organisms. 3rd ed. Cincinnati: USEPA. EPA/600/4-91/002.

[USEPA] U.S. Environmental Protection Agency. 1994b. Short-term methods to estimate the chronic toxicity of effluents and receiving waters to estuarine and marine organisms. 2nd ed. Cincinnati: USEPA Office of Research and Development. EPA/600/4-91/003.

[USEPA] U.S. Environmental Protection Agency. 1994c. Methods for measuring the toxicity and bioaccumulation of sediment-associated contaminants with freshwater invertebrates. Duluth MN: USEPA Office of Research and Development. EPA/600/R-94/024.

[USEPA] U.S. Environmental Protection Agency. 1995. Whole effluent toxicity: guidelines establishing test procedures for the analysis of pollutants. *Federal Register* 60(199):53529–53544.

C. Robert Cappel

Industries in the U.S. are impacted by metals WQC that are derived for the protection of aquatic life either as direct or indirect dischargers. Industries that discharge wastewater directly to receiving waters must have the same permits and comply with the same criteria as municipal wastewater treatment plants. Industries that discharge wastewater to municipal wastewater treatment plants must comply with discharge limits established in pretreatment programs designed to ensure that the treatment plant never violates its permit limits. The use of overly conservative criteria and permitting policies can lead to stringent discharge limits that result in substantial economic costs and increased overall pollution, without any measurable environmental benefits.

Industry's concerns begin with the conservative approaches used in laboratory bioassays to derive WQC. The acute tests are performed over 96 hours using the most toxic forms of the metal in test water that contains minimum complexing agents and nutrients and no dissolved organic carbon, particulates, or sediments. Typically, the organisms are not fed during an acute test. The test conditions alone produce significant stress on the organism without any metal exposure. Control mortality of 10 to 20% is "normal"; LC50 values determined from studies with control mortality even higher than 20% can be found in USEPA's databases. Different life stages are used in different tests, from egg fertilization to embryo-larval to juvenile, with the most stringent toxicity values used to derive the criteria. The following questions deserve consideration:

1) Are the test conditions themselves too conservative to adequately assess acute toxicity in natural waters?
2) Should there be defined concentrations of nutrients and components of natural waters in the test water?
3) Should the organisms be fed during the tests to mimic natural conditions?
4) Should sediment be present in the test chambers?
5) What life stage should be used in the tests?
6) Should chronic or subchronic endpoints be used for acute criteria?

After a sufficient number of bioassays are available, species or genus mean acute LC50 values are derived, which are then assigned arbitrary values of 10, 20, 30, 40, *etc.*, representing the percent of species not protected by those values. A regression is done on these values, plotted, and the graph extrapolated to the 5% level, which determines the final acute value. This value is then divided by 2 to produce the CMC, or acute criterion.

7) Is this manipulation of test data appropriate or overly conservative for deriving acute criteria?

After the acute criteria are established, chronic criteria are derived. This is typically done by deriving an acute:chronic ratio, where the acute LC50 values are compared to the lowest chronic endpoints for each species established in either long-term or short-term chronic tests. The largest ratio, which typically occurs with the most sensitive species, is then used to derive the CCC, or chronic criterion, by dividing the CMC by this ratio.

8) Is this approach overly conservative for deriving the chronic criteria?

The criteria are then expressed as "total recoverable" metal, assuming that all forms are equally toxic. While USEPA has changed their policy to the "dissolved" form, several regional offices and many states do not want to give up the total recoverable concept. One common claim now is that the total recoverable criteria are needed to protect sediments. USEPA's current definition of *dissolved* is the amount passing through a 0.45 micron filter. However, it is common to find another 25 to 50% of the total metal concentration can be removed by filtering through a 0.1 micron filter.

9) Should a strong statement be made in favor of the dissolved form for criteria?

10) Is the 0.45 micron filter definition overly conservative?

11) Should speciation be considered in the criteria?

12) Are colloidal forms of metals toxic?

13) Are dissolved forms less protective of sediment organisms?

More conservative factors are then used in deriving the discharge permit limits. The acute standards are either a maximum allowable concentration or a 1-h average, even though the criteria were derived from 96-h tests. At concentrations around the LC50 values, metals are typically not fast-acting toxicants. The chronic standards are either a 1-d or 4-d average, even though the chronic tests typically range from 28 days to 2 years. Dilution factors are based on the lowest 3- to 7-d flow of the receiving water over the last 2 to 10 years and the design flow of the treatment plant. The most conservative discharge limits are placed on dischargers to dry land.

Mixing zones are supposed to be as small as possible, so some states do not allow any mixing zones. The criteria must be met at the end of the pipe, even when the discharger demonstrates the effluent is not acutely or chronically toxic. Some states have fishable, swimmable, reproducible pipes. The discharger is not allowed to exceed the permit limits by more than 20%, and then only once every three years.

14) What conservative factors should be used in discharge permits and how should these factors be applied?

The use of all the conservative approaches used in deriving the criteria and discharge limits for metals has become a serious problem for both direct and indirect dischargers. The results are increased costs for consumer goods, increased costs for health care, and the use of treatment technologies or alternative discharges that increase the pollution of the environment. Without a scientifically sound, risk-based approach to metals criteria and discharge limits, we end up paying more to pollute more.

A municipal perspective of metals criteria for aquatic life protection

Norman E. LeBlanc

Municipal governments must deal with metal discharges from two distinctively different sources: wastewater treatment plants and storm water. These two types of discharges present different sets of issues that need to be considered at this meeting.

Municipal wastewater treatment plants receive heavy metals from numerous sources and in various forms. Unlike historical conditions, industrial metal inputs are pretreated prior to discharge to the municipal system. Pretreatment modifies the forms of the metals as well as the toxicity. Industrial and domestic metal loadings also undergo further transformations in the anaerobic collection systems and in the aerobic biological treatment process prior to discharge. In general, receiving waters below these discharges contain elevated levels of dissolved and particulate organic matter that can further complex discharged metals. Furthermore, the permitting process sets limits on critical low-dilution periods when effluent concentrations are greatest. The impacts of metal discharges under this scenario cannot be predicted from routine laboratory studies using soluble metal salts dissolved in typical laboratory dilution water that contains low levels of particulate and dissolved organic matter.

Municipal stormwater discharges are unique from wastewater discharges in that they are episodic in nature and generally contain only the more ubiquitous metals (copper, cadmium, lead, zinc, *etc.*). Since these metals originate from such activities as soil erosion or the wearing of tires and brakes, they are usually associated with particulates. Permitting scenarios for these types of discharges are usually identical to those of wastewater discharges. Therefore, prediction of environmental toxicity from the discharge of metals from both wastewater treatment plants and stormwater systems is overestimated by conventional laboratory extrapolations. Informal surveys of laboratories performing municipal TIEs confirms that metals in wastewater discharges are not bioavailable. However, in very isolated cases where industrial discharges of metals are the dominant loadings to the treatment system, can metal toxicity be manifested in treated effluents?

There are several outstanding questions that do need to be addressed. These include bioavailability issues of fate and transport, documentation of demonstrated receiving system impacts caused by metals, the separation of historical from ongoing problems, as well as innovative treatment solutions for the identified problem areas.

Metals pose serious management concerns to municipal governments, given the ubiquitous nature of many metals coupled with the lack of demonstrated environmental impacts associated with current activities. Conflicting priorities between dwindling resources and competing interests such as schools and public safety necessitate that local governments ensure that real environmental benefits accrue from targeted activities.

Review of international regulatory practices and alternative hazard identification scheme for metals and metal compounds

Guy Ethier

The Intergovernmental Forum on Chemical Safety (IFCS), under the direction of the United Nations Conference on Environment and Development (UNCED), has the responsibility to develop a harmonized approach for the hazard identification/classification and labeling of chemicals. This initiative includes the class "Dangerous to the aquatic environment." The main classification systems for hazard to aquatic organisms, which the harmonization addresses, are 1) the IMO/GESAMP profile for marine pollutants, 2) the EU supply and use system, and 3) the ADR/RID system for transport on road and rail. The Government of Canada has a policy that assesses substances and recommends risk management approaches based on level of hazard.

Currently, the EU and the Government of Canada have formalized hazard identification approaches for the aquatic environment that take into account toxicity, persistence and bioaccumulation, and anthropogenic origin. These criteria are applied indiscriminately to inorganic and organic compounds, irrespective of form or physical association.

Over the last year, the two international scientific workshops were held. The first, "Aquatic Toxicity Testing of Sparingly Soluble Metals, Inorganic Metal Compounds and Minerals," was sponsored by the EU and Canada and held in Brussels in January 1996. The second, "Biodegradation/Persistence and Bioaccumulation/Biomagnification of Metals and Metal Compounds," was sponsored by OECD and held in Ottawa, Canada, in September 1995.

The unique properties of metals compared to organic substances and the agreements reached at the two workshops have lead to the development of a proposed alternative for a hazard identification scheme that recognizes the particular properties of metals and metal compounds. The proposed alternative is based on three components: toxicity, transformation and solubilization, and biomagnification.

While addressing the above, the following key points and tables will also identify areas of international consensus and areas that need further work in order to render the proposed alternative viable and useful.

International Harmonization
- UNCED Agenda 21, Chapter 19 six major areas for work
  Program Area B: Harmonization for Classification and Labelling
- "Share the Burden"
  OECD (though EC and BIAC) responsible for proposal on "Aquatic Toxicity Criteria and Classification Systems" (target date: 1997)
- IMO/GESAMP profile for marine pollutants
- ADR/RID system for transport by road or rail
- The Canadian Toxic Management Policy
- US-DOT, Pesticides and Industrial Chemicals
- EEC Directive on Supply and Use

✓ "Dangerous to the environment" classification based on: persistence, bioaccumulation/biomagnification, toxicity

✓ Organics and inorganics (*e.g.*, metals) treated similarly

Some Problems

- Organics and inorganics are intrinsically different
- Persistence not a useful discriminator for naturally occurring elements
- Biomagnification is predicted, incorrectly, from bioaccumulation

Current Efforts To Address Problems For Sparingly Soluble Inorganic Materials (SSIMs)

- Ongoing international discussion and dialogue
  OECD Advisory Group on Harmonization of Classification
- OECD Workshop (Ottawa, Canada, September 1995)
  "Aquatic toxicity testing of sparingly soluble metals, inorganic metal compounds and minerals"
- Canada/EU Workshop (Brussels, Belgium, December 1995)
  "Technical Workshop on biodegradation/persistence and bioaccumulation/ biomagnification of metals and metal compounds"

International Consensus From OECD Workshop Ottawa

- Bioavailability key to hazard identification
- Hazard begins with acute toxicity
- Long-term (chronic) toxicity needs to be considered

Major Issues from OECD Workshop Ottawa

- Agreed: testing only to solubility limits
- Agreed: no use of vehicles or additives
- No agreement: how to report results for hazard identification *in terms of the total concentration of material tested or in terms of only the soluble fraction*

Consensus from Brussels Workshop

- Agreed: Bioavailability is key to hazard identification (also from Ottawa Workshop)
- Agreed: Persistence is a secondary measure; should not be used alone for hazard identification and should be interpreted in terms of chemical and biological transformation

- Agreed: Bioaccumulation is not useful for hazard identification but is useful for risk assessment

Objective of Hazard Identification

*to recognize the **potential** of a substance to cause adverse effects in the environment under conditions of both transport (including releases such as spills) and supply (release during usage)*

Principles Underlying the Proposed Approach

- Hazard identification and risk assessment differ (presently confused by regulators and others)
- Hazard identification should only be used for informational labeling and possibly for priority setting
- Management decisions should be based on risk assessments
- The only intrinsic property is toxicity (acute and chronic)

A Realistic Hazard Identification Approach for SSIMs

note:　✓ Building on results from the Ottawa and Brussels Workshops,

　　　　✓ "Fitting" within the proposed OECD framework

- Key is bioavailability measured by toxicity (acute and chronic)
- Persistence is related to transformation into more bioavailable forms
- Bioaccumulation is not useful for hazard identification
- Biomagnification needs to be assessed on a case by case basis

Hazard identification components for organics and SSIMs[1]

Types of substances	Hazard identification components		
	Toxicity	Persistence (= bioavailability)	Bioaccumulation (= long-term effects and biomagnification)
Organics	EC_x in mg/L	Biodegradability	BCF, log K_{ow}
SSIMs[1]	EC_x in mg/L [2]	Chemical and biological transformation[3]	Biomagnification (Long-term effects evaluated by chronic toxicity testing)

[1] SSIMs = Sparingly soluble inorganic materials; includes metals, inorganic metal compounds and minerals.
[2] Includes both acute and chronic (*i.e.*, long-term) toxicity.
[3] Equilibrium solubility of sparingly soluble inorganic materials (K_{sp}) and the rate of solubilization.

Endpoints and evaluation factors for hazard identification of SSIMs

Component	Endpoints	Evaluation factors
Toxicity		
Acute (short-term)	EC50 in mg/L	Marine and freshwater tested
Chronic (long-term)	$EC_{chronic}$ in mg/L	separately: fish, crustacea, algae
Bioavailability		
Chemical transformation	Equilibrium solubility and rate	Moderating factors for toxicity[1]
Biological transformation	of solubilization[1]	Available information evaluated
	Case-by-case evaluation	by expert judgment
Biomagnification	Case-by-case evaluation	Available information evaluated by expert judgment

[1] Value used should be some proportional indication of long-term entry into the environment.

Advantages and disadvantages of the different options for reporting SSIMs toxicity test data for hazard identification

	Option 1: total concentration[1]	Option 2: soluble fraction of evaluated compound and EC_x of a representative soluble species[1]	Option 3: soluble fraction of evaluated compound and EC_x of the same compound[1]
Advantages	• Evaluated compound tested directly • Toxicity and bioavailability[2] directly integrated • Speciation taken into account directly • Appropriate for mixtures and alloys	• Less testing required	• Evaluated compound tested directly • Speciation taken into account directly • Appropriate for mixtures and alloys
Disadvantages	• More testing required	• Evaluated compound not tested directly • Toxicity and bioavailability not directly integrated • Speciation not taken into account directly • Not appropriate for mixtures and alloys	• Toxicity and bioavailability not directly integrated • Most testing required

[1] Toxicity endpoints (EC_xs) expressed as the dissolved fraction of the test compound, in mg/L.
[2] Bioavailability = chemical and biological transformation.

Proposed OECD hazard categories

Degradation	Toxicity (LEC 50 mg/l)	Bioaccumulation	Conditions for Classification
If negative	< 1.0	If positive	Box 2
Box 1	Box 2	Box 3	Boxes 2 & 1
			Boxes 2 & 3
If negative	>1.0 <10.	If positive	Boxes 4 & 1
Box 1	Box 4	Box 3	Boxes 4 & 3
If negative	>10.0 <100.0	If positive	Boxes 5 & 1
Box 1	Box 5	Box 3	
	SAFETY NET		
If negative	>10.0 <100.0	If positive	Boxes 1 & 3
Box 1	Box 5	Box 3	

Summary table of different options for SSIMs

Substance	Toxicity assessment	Persistence	Bioaccumulation/ biomagnification
Organic (existing approach)	Highly, moderately, slightly, or non-toxic relative to possible "dangerous for the environment" (DFE) classification	Biodegradation study required; substances which biodegrade may not be labeled DFE, non-biodegradable substances may be labeled persistent	Log k_{ow} or measured BCF; may be used to label a substance bioaccumulative as well as toxic
SSIMs (option 1)	Same assessment, but based on "total" SSIM used to prepare the dose	Assess transformation which may increase bioavailability	Assess biomagnification based on available data (literature) and professional judgment
SSIMs (option 2)	Same assessment, but tests are performed with one soluble salt for each SSIM with extrapolation to calculate the amount needed to produce an EC50 for other forms of the same SSIM; solubility factor required for each metal	Assess transformation which may increase bioavailability (solubility factor for chemical transformation)	Assess biomagnification based on available data (literature) and professional judgment
SSIMs (option 3)	Same assessment, but tests are performed on each product sold and all EC50 are extrapolated to "total dose" via a solubility factor for each product	Assess transformation which may increase bioavailability (solubility factor for chemical transformation)	Assess biomagnification based on available data (literature) and professional judgment

Proposed Framework for Hazard Identification of SSIMs

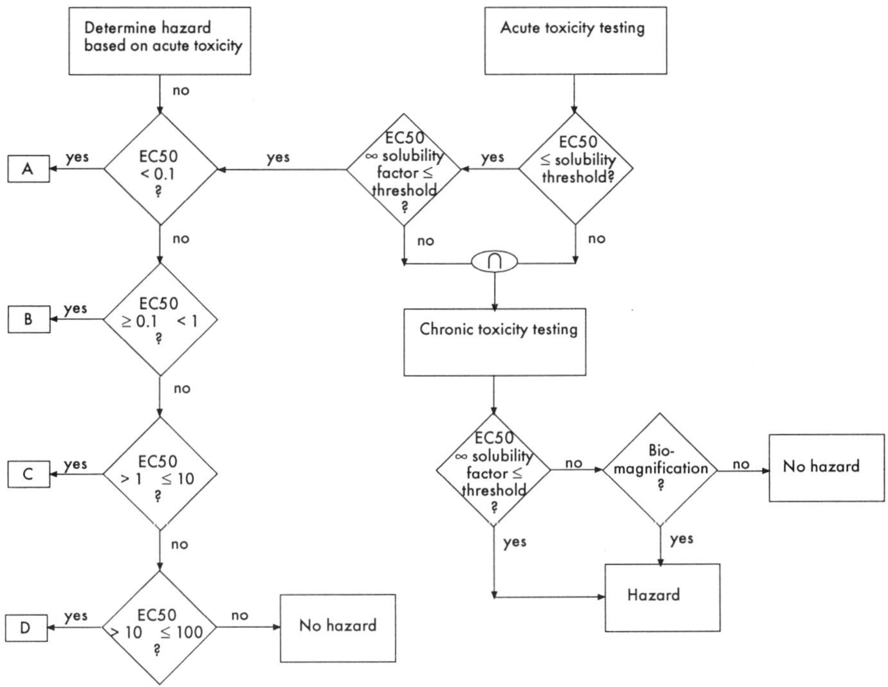

Possible outcomes of proposed hazard identification framework

Hazard assessment components:		Possible outcomes											
Biomagnification		No						Yes					
Chronic toxicity		Yes			No			Yes			No		
Solubility factor		High	Moderate	Low	High	Moderate	Low	High	Moderate	Low	High	Moderate	Low
Acute toxicity	High	H	H	H	H	H	(U)	H					
	Moderate	H	H	(U)	H	(U)	NH						
	Low	H	(U)	(U)	(U)	NH	NH						
	None	(U)	(U)	NH	NH	NH	NH				NH		

H = Hazard; NH = No hazard; (U) = uncertain; depends on testing outcomes and thresholds

Critical Research Needs
- General dissolution protocol for obtaining the relevant soluble fraction of an SSIM *from the Ottawa Workshop*
- Standardized protocol for determining chemical solubility factors for SSIMs *from the Brussels Workshop*
- Testing/validation of the proposed framework
- Valuable information re: Biological transformation of SSIMs

European Union Ecotoxicity Classification and Labeling Requirements

"Dangerous for the environment"

    R50:   Very toxic to aquatic organisms
             < 1 mg/liter [1]

    R51:   Toxic to aquatic organisms
             1 to < 10 mg/liter [1]

    R52:   Harmful to aquatic organisms
             10 to < 100 mg/liter [1]

    R53:   May cause long-term adverse effects in aquatic organisms
             $\log P_{ow} > 3$ or BCF > 100

[1] IC_{50} Algae 72h or EC_{50} Daphnia 48h or LC_{50} Fish 96h

Acknowledgments

Thanks to Dr. Peter Chapman who prepared the manuscript. Thanks to the members of the harmonization working group:

     Robert Prairie, Noranda

     Hudson Bates/Larry Cursio, NiPERA

     Craig Boreiko, ILZRO

     Brigitte Dero, Eurométaux

     Elaine Dorward-King /Bill Adams, Kennecott Copper Utah

     Rick Hilton, INCO

     Larry Morris, Falconbridge

     Hugo Waeterschoot, Union Minière

     Justyna Laurie-Lean, Mining Association Of Canada

     Kristina Ringwood, Western Mining Corp.

     Walter Kuit, Cominco

     Bruce Conard, INCO

     Christopher Lee, ICA

Physiological and toxicological effects of metals at gills of freshwater fish

Richard C. Playle

Gills of freshwater fish have two important physiological functions. First, they are the site of gas transfer (O_2, CO_2, NH_3) for the fish. Second, in freshwater fish, they are the site of active ion uptake (Na, Cl, Ca) to counter diffusive loss of ions down their electrochemical gradients from the fish to the water (Wood 1992). Fish gills are the first organ affected by many waterborne metals. At low concentrations, metals that affect gill function usually do so through ionoregulatory disturbances and cause respiratory distress only at relatively high concentrations.

In my talk, I will present data of mine and others indicating the physiological effects of metals such as aluminum, copper, cadmium, and silver. Aluminum is an interesting metal because it causes respiratory effects even at low, environmentally relevant concentrations. These respiratory effects occur as moderately acidic water containing aluminum passes over the gills and becomes more alkaline, and aluminum polymerizes and precipitates onto the gills (Playle and Wood 1989b). The gill microenvironment is made more basic because NH_3 is released at the gills (Neville 1985; Playle and Wood 1989a; Lin and Randall 1990).

Basic water is made more acidic as it passes over the gills because CO_2 is released at the gills (Lloyd and Herbert 1960; Playle and Wood 1989a; Lin and Randall 1990). Carbonic anhydrase is thought to catalyze CO_2 hydration at the gills, otherwise CO_2 would not dissociate fast enough to acidify water near the gills.

In the gill microenvironment, pH changes may sometimes be important when modeling metal speciation in water. If the water of interest has low buffer capacity, then pH changes near the gills may be important if the metal considered has important shifts in species or solubility as water pH changes (Playle *et al.* 1992).

As for the other metals I will mention, copper affects active Na^+ uptake at the gills (Laurén and McDonald 1985), cadmium affects active Ca^{2+} uptake at the gills (Verbost *et al.* 1989), and silver may interfere with Na^+ uptake (Wood *et al.* 1995). Toxicities of these metals are all reduced in water with higher Ca^{2+} and DOC concentrations, and H^+ ions usually protect against these metals (but are toxic themselves). A good conceptual model for these protective effects is the "gill surface interaction model" presented by Pagenkopf (1983) and by Morel (1983) and Morel and Hering (1993).

The gill surface interaction model of Pagenkopf (points 1–6) and Morel (point 7) is paraphrased below:

1) Acutely toxic metals alter gill function.
2) Certain species of metals are more toxic than others.
3) Metals and H^+ form complexes with negatively charged gill surfaces.
4) Rates of metal-gill interactions are fast.
5) Gill surfaces have a finite interaction capacity.

6) Competition occurs between water hardness cations, metals, and H^+.

7) Availability of free metal ions is reduced through complexation with aquatic ligands such as dissolved organic carbon.

During my presentation I will cover most of these points, mainly using silver as an example. The current approach in my laboratory is to measure metal deposition on fish gills after exposure of the fish to the metal in a synthetic softwater system. The water is then modified by adding cations such as Ca^{2+}, Na^+, and H^+ (competition experiments, point 6) or various ligands such as EDTA, thiosulphate, or natural DOC (complexation experiments, point 7). By running these types of experiments using synthetic ligands of known metal binding strength, a metal-gill conditional stability constant can be calculated (the "ligand exchange" method).

For example, thiosulphate ($S_2O_3^=$) binds silver relatively strongly (log $K_{Ag-S_2O_3}$=8.8). From our experiments, we determined that log $K_{Ag-gill}$=10.0 and log K_{Ag-DOC}=9.0 (Janes and Playle 1995). Once determined, these values can be entered into an aquatic chemistry computer program such as MINEQL$^+$ (Schecher and McAvoy 1992) or MINTEQA2/ PRODEFA2 (Allison *et al.* 1991).

In essence, we inserted biology (=fish gill) into a powerful aquatic chemistry program, MINEQL$^+$, to better predict interactions of metals such as silver with sensitive biological membranes. This approach to modeling biological surfaces is theoretically an extremely valuable tool. Modeling metal-gill interactions has been done for copper and cadmium (Playle *et al.* 1993a, 1993b) and lead, cobalt, and mercury (Richards, Kuehn, Hughes, and Playle, unpublished). Generally the metals bind to the gills according to their strength of binding to synthetic ligands. Exceptions to the expected binding order are interesting in that they reflect physiological mechanisms at the gills. For example, the better than expected binding of cadmium to the gills (Playle *et al.* 1993b) probably reflects active uptake of cadmium at high affinity calcium channels (Verbost *et al.* 1989).

Although this approach to modeling is useful, it is not without its practical problems. For example, Shuttleworth and Unz (1991) tried to model the effects of calcium, copper, nickel, and zinc on growth of filamentous algae. They listed some analytical and modeling problems they encountered when modeling their results using MINEQL. These problems were 1) unknown equilibrium constants for components in their system; 2) the opposite problem, more than one equilibrium constant for a reaction, with no indication of which value might be better to use; 3) reaction kinetics are not easily considered; 4) the copper ion selective electrode they used ran into some interferences, as did nickel and zinc measurements; and 5) H^+ ion concentration had a large effect on their modeling results, and their pH measurements were accurate only to ± 0.02 pH units. Still, the agreement between measured and predicted values was good.

The value of a binding "constant" is dependent on experimental conditions used to derive that value, especially the concentration of metal used. Exposing a fish gill (or algae or DOC or humic acid) to a low metal concentration will fill only a few, high affinity sites such as sulphydryl groups (*e.g.*, average log $K \sim 9.5$). At high metal concentration, more sites of lower and lower affinity are filled, so that the average binding "constant" is much lower (*e.g.*, average log K~6.0).

Complex molecules composed of humic and fulvic acids, such as DOC, do not have a discrete equilibrium binding constant but some average value with a distribution about the mean value (above). Perdue and Lytle (1983) used a Gaussian distribution model to describe complex ligand mixtures such as humic acids. Their distribution used the mean log K value of the ligand mixture plus the statistical distribution about this mean value. Grimm *et al.* (1991) described a similar, continuous multiligand distribution for copper binding to Suwannee River dissolved organic matter.

There is some debate about the general applicability of binding constants determined in one study to other situations, especially with complex molecules such as DOC and fulvic and humic acids. However, Cabaniss and Shuman (1988a) used Suwannee River fulvic acid (FA) and showed that Cu:FA binding was > 90% 1:1 binding, that H^+ competition was important but Ca^{2+} and Mg^{2+} competition was not, and that ionic strength had effects on Cu-FA binding. In a companion paper, they compared different sources of organic matter and concluded that FA and DOC differences are less important than are pH and ionic strength effects (Cabaniss and Shuman 1988b). This conclusion differs from that obtained from the compilation of Buffle (1984), which indicated 2 to 3 log unit differences due to DOC source. Cabiniss and Shuman suggested that different methodologies were the largest contributors to this variability. In agreement with the general applicability of DOC results, Sahu and Banerjee (1990) showed that soil and peat humic acids varied in their metal binding by <0.5 log units, and, using a non-metal example, log $K_{DOC\text{-}dieldrin}$ values were the same for sediments from three Minnesota lakes (Kosian *et al.* 1995).

From my own viewpoint, I agree with the conditional nature of equilibrium binding constants but feel that if models are constructed to be applicable 1) at environmentally low metal concentrations (*e.g.*, 0.1 to 1.0 μM), 2) at reasonable pHs (*e.g.*, pH 5 to 9), and 3) for reasonable DOC concentrations (*e.g.*, 0 to 10 mg CL^{-1}), then the models will be useful for predictive purposes.

Such models are extensions of empirical models such as those of Meador (1991) and Welsh *et al.* (1993) that described toxicity of copper to *Daphnia magna* and fathead minnows, respectively, as variables of total copper concentration, water pH, and DOC concentration. What is probably needed in the future is to use metal-gill interaction models to predict acute toxicities while correlating gill metal concentrations with observed mortalities. This approach has been used in a one-compartment, first-order kinetic pulse-exposure model for whole body residues and lethality (Hickie *et al.* 1995). The strength of gill interaction models comes from the power behind the computer programs themselves and because interactions at the gills, the target organ of the metals, are directly considered.

References

Allison JD, Brown DS, Novo-Gradac KJ. 1991. MINTEQA2/PRODEFA2, a geochemical assessment model for environmental systems: Version 3.0 User's manual. Washington DC: USEPA.

Buffle J. 1984. Natural organic matter and metal-organic interactions in aquatic systems. In: Siegel H, editor. Metal ions in biological systems. New York: Marcel Dekker. p 165–221.

Cabaniss SE, Shuman MS. 1988a. Copper binding by dissolved organic matter: I. Suwannee River fulvic acid equilibria. *Geochim Cosmochim Acta* 52:185–193.

Cabaniss SE, Shuman MS. 1988b. Copper binding by dissolved organic matter: II. Variation in type and source of organic matter. *Geochim Cosmochim Acta* 52:195–200.

Grimm DM, Azarraga LV, Carreira LA, Susetyo W. 1991. Continuous multiligand model used to predict the stability constant of Cu(II) metal complexation with humic material from fluorescence quenching data. *Environ Sci Technol* 25:1427–1431.

Hickie BE, McCarty LS, Dixon DG. 1995. A residue-based toxicokinetic model for pulse-exposure toxicity in aquatic systems. *Environ Toxicol Chem* 14(12):2187–2197.

Janes N, Playle RC. 1995. Modeling silver binding to gills of rainbow trout (*Oncorhynchus mykiss*). *Environ Toxicol Chem* 14(11):1847–1858.

Kosian PA, Hoke RA, Ankley GT, Vandermeiden FM. 1995. Determination of dieldrin binding to dissolved organic material in sediment pore water using a reverse-phase separation technique. *Environ Toxicol Chem* 14(3):445–450.

Laurén DJ, McDonald DG. 1985. Effects of copper on branchial ionoregulation in the rainbow trout, *Salmo gairdneri* Richardson. *J Comp Physiol B Metab Transp Funct* 155:635–644.

Lin H, Randall DJ. 1990. The effect of varying water pH on the acidification of expired water in rainbow trout. *J Exp Biol* 149:149–1 60.

Lloyd R, Herbert DWM. 1960. The influence of carbon dioxide on the toxicity of un-ionized ammonia to rainbow trout *(Salmo gairdneri* Richardson). *Ann Appl Biol* 48:399–404.

Meador JP. 1991. The interaction of pH, dissolved organic carbon, and total copper in the determination of ionic copper and toxicity. *Aquat Toxicol* 19:13–32.

Morel FMM. 1983. Principles of aquatic chemistry. Toronto ON: J Wiley. 446 p.

Morel FMM, Hering JG. 1993. Principles and applications of aquatic chemistry. New York: J Wiley. 588 p.

Neville CM. 1985. Physiological response of juvenile rainbow trout, *Salmo gairdneri*, to acid and aluminum: prediction of field responses from laboratory data. *Can J Fish Aquat Sci* 42:2004–2019.

Pagenkopf GK. 1983. Gill surface interaction model for trace-metal toxicity to fishes: role of complexation, pH, and water hardness. *Environ Sci Technol* 17:342–347.

Perdue EM, Lytle CR. 1983. Distribution model for binding of protons and metal ions by humic substances. *Environ Sci Technol* 17:654–660.

Playle RC, Dixon DG, Burnison K. 1993a. Copper and cadmium binding to fish gills: modification by dissolved organic carbon and synthetic ligands. *Can J Fish Aquat Sci* 50:2667–2677.

Playle RC, Dixon DG, Burnison K. 1993b. Copper and cadmium binding to fish gills: estimates of metal-gill stability constants and modelling of metal accumulation. *Can J Fish Aquat Sci* 50:2678–2687.

Playle RC, Gensemer RW, Dixon DG. 1992. Copper accumulation on gills of fathead minnows: influence of water hardness, complexation and pH of the gill micro-environment. *Environ Toxicol Chem* 11(3):381–391.

Playle RC, Wood CM. 1989b. Water pH and aluminum chemistry in the gill micro-environment of rainbow trout during acid and aluminum exposures. *J Comp Physiol B Metab Transp Funct* 159: 539–550.

Playle RC, Wood CM. 1989a. Water chemistry changes in the gill micro-environment of rainbow trout: experimental observations and theory. *J Comp Physiol B Metab Transp Funct* 159:527–537.

Sahu S, Banerjee DK. 1990. Complexation properties of typical soil and peat humic acids with copper(II) and cadmium(II). *Int J Environ Anal Chem* 42:35–44.

Schecher WD, McAvoy DC. 1992. MINEQL+: A software environment for chemical equilibrium modeling. *Computers, Environment and Urban Systems* 16:65–76.

Shuttleworth KL, Unz RF. 1991. Influence of metals and metal speciation on the growth of filamentous bacteria. *Water Res* 25:1177–1186.

Verbost PM, Van Rooij J, Flik G, Lock RAC, Wendelaar Bonga SE. 1989. The movement of cadmium through freshwater trout branchial epithelium and its interference with calcium transport. *J Exp Biol* 145:185–197.

Welsh PG, Skidmore JF, Spry DJ, Dixon DG, Hodson PV, Hutchinson NJ, Hickie BE. 1993. Effect of pH and dissolved organic carbon on the toxicity of copper to larval fathead minnow *(Pimephales promelas)* in natural lake waters of low alkalinity. *Can J Fish Aquat Sci* 50:1356–1362.

Wood CM, Morgan I, Galvez F, Hogstrand C. 1996. The toxicity of silver in fresh and marine waters. In: Andren AW, Bober TW, editors. Third International Conference Proceedings: Transport, Fate and Effects of Silver in the Environment; 1995 Aug 6–9; Washington DC. Madison: Univ Wisconsin. p 51–59.

Wood CM. 1992. Flux measurements as indices of H+ and metal effects on freshwater fish. *Aquat Toxicol* 22:239-264.

Modeling the interactions of metals with dissolved organic matter in surface waters

Edward Tipping

The last thirty years have seen numerous laboratory studies of metal and proton binding by humic substances, aimed both at elucidating mechanisms and providing quantitative information. Recent modeling studies have begun to make more direct use of the results in order to try to explain and predict metal behavior in natural waters. Topics of interest have included aluminium speciation in acid surface waters, heavy metal chemistry in soils, radionuclides in ground waters, and toxic metals in surface waters.

Humic Ion-Binding Model V was developed specifically to make use of available laboratory data on proton and metal binding by humic substances. An effort was made to minimize the number of model parameters in order that as many datasets as possible, however small, could be analyzed to obtain model parameters. The aim was to obtain an overall description of average humic behavior. The model is based on accepted physico-chemical mechanisms of ion-humic interactions. The model has been described in detail, including the full algebra, in previous publications (Tipping and Hurley 1992; Tipping 1993a, 1993b). Its essential features are as follows:

- Humic compounds are represented by hypothetical size-homogeneous molecules, which carry proton-dissociating groups that can bind metal ions. The interactions are described in terms of intrinsic equilibrium constants—which refer to the (usually hypothetical) situation where the humic substances have zero electrical charge—and an electrostatic term, which takes into account the influence on binding of the variable humic charge.

- The proton-binding groups of the humic substances are heterogeneous, having a range of pK_{intr} values. Two types of acid group are distinguished, denoted by A and B. Within each type there are four different groups present in equal amounts, the pK values of which are described in terms of a median value, pK_A or pK_B, and a factor, ΔpK_A or ΔpK_B, that defines the range of the values.

- Metal binding takes place at single proton-binding sites (monodentate) and at bidentate sites formed by pairs of proton-dissociating sites. The extent to which bidentate binding can take place is constrained by the proximity factor (f_{pr}), which defines, on the basis of molecular geometry, the likelihood of pairs of proton-binding groups being close enough to form bidentate sites.

- Monodentate metal binding is described with intrinsic equilibrium constants for the metal-proton exchange reaction

$$RAH^Z + M^{z+} = RAM^{Z+z} + H^+.$$

The negative logarithms of these intrinsic constants are denoted by pK_{MHA} and pK_{MHB} for type A and type B sites respectively. The appropriate pK_{MH} values are simply added together to describe bidentate binding. The model permits the binding of the first hydrolysis product (*e.g.*, CuOH$^+$ in the case of Cu^{2+}) as well as the parent species. The pK_{MHA} value for the first hydrolysis product is assumed to be the same as that of the parent. The smaller the value of pK_{MHA}, the stronger the binding of the metal species in question.

- The (nonspecific) accumulation of an excess of counterions in the diffuse layer adjacent to the molecular surface also contributes to the total binding and is described with simple Donnan expressions.

Parameter estimation can be considered in two parts. First, the dissociation of protons has to be characterized (6 parameters). Second, metal binding is described in terms of metal-proton competition; in the basic model, pK_{MHA} and pK_{MHB} can be used, but the two pK_{MH} values are correlated, so that often a single parameter (pK_{MHA}) suffices to fit data. Parameter values have been derived from experimental data for 20 metals, and other values have been estimated from linear free-energy relationships.

The WHAM (Tipping 1994) is an equilibrium computer code in which Model V is combined with a model for inorganic solution speciation and with other sub-models that pertain mostly to soil systems. WHAM allows speciation calculations to be done for natural waters, given the total concentrations of reacting master species, including fulvic and humic acids.

Application to natural waters. Model V has been reasonably successful in describing and rationalizing laboratory data, but it is perhaps more important to know how well it can predict humic binding behavior in the field. Data for such testing are sparse, but some examples include these:

- Speciation of cadmium spiked into river waters (Gardiner 1974)
- Aluminium speciation in acid waters
- Charge-(im)balance in acid waters
- Trace copper speciation in lakes (Xue and Sigg 1993)
- Speciation of lead spiked into river waters
- Speciation of copper spiked into bog waters (Dwane and Tipping 1996, unpublished)

Application of WHAM to these data has been done by fixing all model parameters and adjusting the concentration of "active" fulvic acid to get the best fit. It has been found that between 30% and 100% of the dissolved organic matter behaves as would isolated fulvic acid.

Model VI, an enhancement of Model V, improves the description of metal binding at trace levels. Model VI includes a fixed parameter that allows greater heterogeneity of binding sites.

Further research needs. There are several topics that need to be addressed to improve and test Model V, Model VI, or any other humic ion-binding model. At the fundamental level, information is needed on metal binding at high pH, competition effects among metals, and the possible binding of complexed metals, *e.g.*, carbonate complexes. In addition, more field data are needed, obtained specifically to test predictive models.

Acknowledgments. Much of the development of Model V has been funded by Her Majesty's Inspectorate of Pollution, Department of the Environment. Additional support has come from the Natural Environment Research Council.

References

Dwane G, Tipping E. 1996. (unpublished).

Gardiner J. 1974. The chemistry of cadmium in natural water: I. A study of cadmium complex formation using the cadmium specific-ion electrode. *Water Res* 8:23–30.

Tipping E, Hurley MA. 1992. A unifying model of cation binding by humic substances. *Geochim Cosmochim Acta* 56:3627–3641.

Tipping E. 1994. WHAM - A chemical equilibrium model and computer code for waters, sediments and soils incorporating a discrete-site / electrostatic model of ion-binding by humic substances. *Computers & Geoscience* 39:505.

Tipping E. 1993a. Modeling the competition between alkaline earth cations and trace metal species for binding by humic substances. *Environ Sci Technol* 27:520–529.

Tipping E. 1993b. Modeling ion binding by humic acids. *Colloids Surf* 73:113–171.

Xue H, Sigg L. 1993. Free cupric ion concentration and Cu(II) speciation in a eutrophic lake. *Limnol Oceanogr* 38:1200–1213.

Modeling the environmental fate and transport of metals

John P. Connolly

Mathematical representations (models) of the fate and transport of metals in aquatic environments are developed typically as a means to predict the relationship between metals sources and biota exposure concentrations. The quantitative expression of this relationship constitutes the model and it can be used define to a maximum allowable discharge to the environment or to estimate the benefits derived by remediation of a source.

A convenient way to view the fate and transport model is in terms of five component parts:

1) hydrodynamics
2) sediment transport
3) sorption
4) speciation
5) precipitation/dissolution

Each component has been the subject of extensive scientific investigation that has allowed the development of sophisticated models. However, the informational requirements of these models are formidable and their computational requirements can be extreme. Consequently, aggregation of these models into a fate and transport model remains beyond the current state-of-the-art. Informational and computational constraints are particularly relevant for the routine application of models for WLA and TMDL studies. Thus, fate and transport models have tended to use relatively simplistic descriptions of some or all of the components.

Numerous examples exist of the practical application of fate and transport models. Riverine WLAs are generally conducted using an SD approach. Ambient concentration at the point downstream where the effluent is mixed over the cross-section is computed at maximum metal discharge and a statistically defined minimum stream flow. This approach has been refined by using Monte Carlo sampling of statistical characterizations of flows, metals concentration, and hardness to generate a frequency distribution of instream concentrations for comparison to WQC. Although most applications of this PD approach have considered only total recoverable metal, an equilibrium description of sorption has recently been included, allowing the computation of instream dissolved metal concentration. Figure 1 presents distributions of instream total and dissolved silver concentrations generated in this manner. The upper panel shows the distribution of daily average concentrations and the bottom panel shows the distribution of 4-d averages. These distributions are interpreted relative to the statistical expression of the criterion. For example, a value expected to be exceeded once in three years is the 99.9 percentile value $[100 \times (1 - 1/(3 \times 365))]$ of the daily average distribution (15 µg dissolved silver/l) or the 99.6 percentile value $[100 \times (1 - 1/(3 \times 365/4))]$ of the 4-d average distribution (5 µg dissolved silver/l).

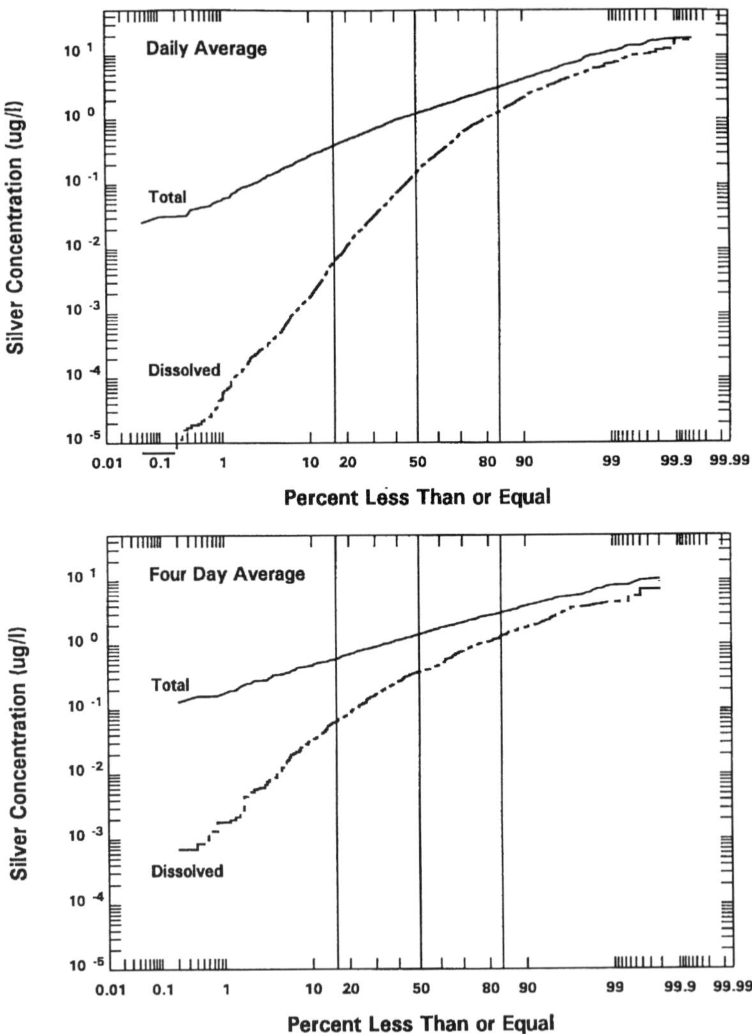

Figure 1 Computed total and dissolved silver concentrations

When multiple metals sources and/or sediments are at issue, models that resolve the spatial component of the problem are required. These models typically assume a steady-state condition in the manner of biological oxygen demand (BOD) WLA models. Thus, a simplistic representation of hydrodynamics and sediment transport is invoked for some defined low-flow condition. Further, these models have not considered speciation or precipitation, except through empirically derived partition coefficients and hardness corrections in the water column and representation of precipitation in sediments by specification of high-sorption partition coefficients. Despite these simplifications, these models can provide reasonable predictions of water column metal concentrations as shown by applications to the Genesee River (HydroQual, Inc. 1982), the Blackstone River (Limno-Tech, Inc. [LTI] 1993) and New York/New Jersey Harbor (HydroQual, Inc. 1995b). An unresolved issue with these models is their capability to describe metal contamination in sediments. For example, Figure 2 shows the Blackstone River copper calibration for a low-flow period in August 1991. The upper two panels demonstrate that the model is capable of reasonable simulation of water column total and dissolved copper. The bottom panel shows that the model is less capable of predicting the sediment copper concentration.

The relationship between metal sources and metal concentrations in sediments has been the subject of recent fate and transport modeling efforts. These efforts have been directed to two major aspects of the problem: 1) the transport of metals between the sediment and water column and 2) the fate of metals within the sediment. Two examples are cited; the first is a zinc problem in the Pawtuxet River, which is addressed using a time variable model that incorporates state-of-the-art representations of hydrodynamics and sediment transport and explicit consideration of zinc-sulfide equilibrium chemistry (HydroQual, Inc. 1995a); the second is a cadmium problem in which the kinetics of sulfide oxidation and metal release to the water column are examined (Di Toro *et al.* submitted).

An interesting aspect of the Pawtuxet River application is the sensitivity of computed sediment metals concentrations to the variable hydrograph. This is illustrated by the following numerical experiment. The surficial sediment zinc concentration in steady-state with a constant mass loading of zinc to the river was computed for two cases: 1) constant low-flow typical of the condition simulated in a steady-state WLA model and 2) the variable hydrograph characteristic of the river (Figure 3). In the first case, the sediment concentrations changed little from the concentrations specified at the beginning of the ten-year simulation. In the second case, the sediment concentrations changed dramatically from the initial conditions. Little change occurred at the low-flow condition because of the lack of interaction between the water and sediment. No resuspension was occurring, settling was minimal because of the low water column suspended solids, and little diffusion occurred because of the low concentration gradient between the water and sediment. In contrast, the high flows occurring in the variable hydrograph resulted in both significant resuspension and settling and much greater flux of zinc between the water and sediment. The importance of flood events to the level of sediment contamination sug-

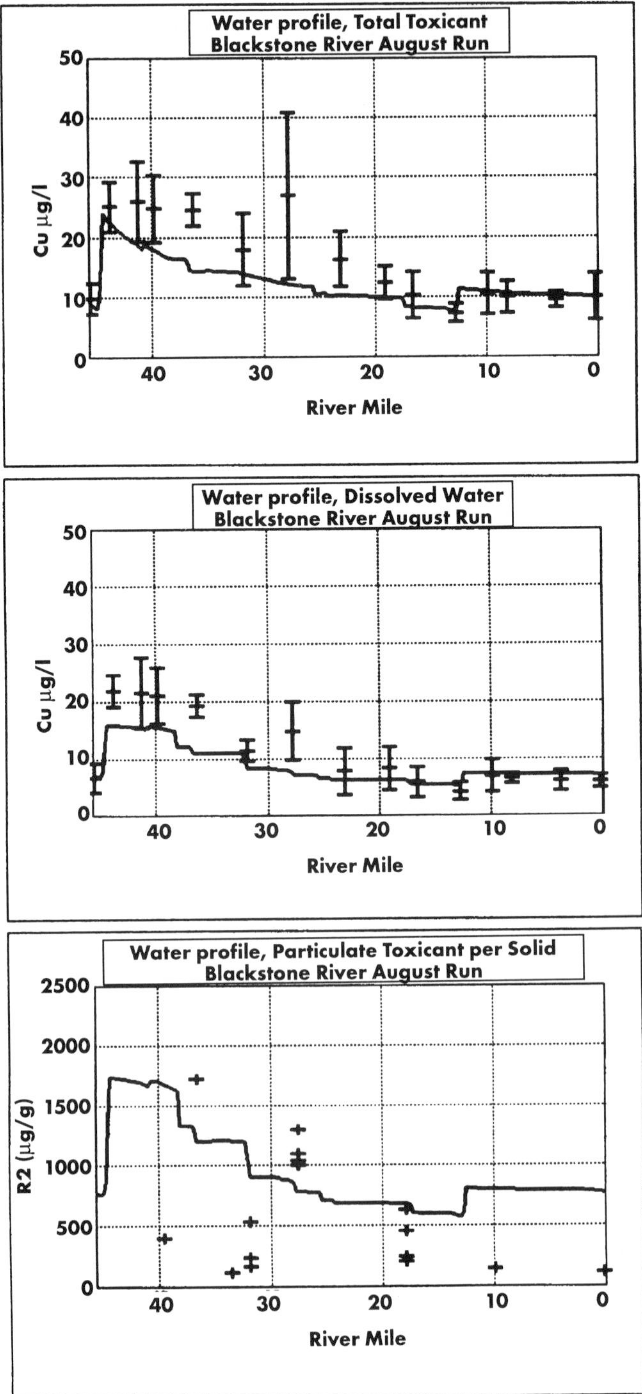

Figure 2   Total, dissolved, and particulate toxicants in water and bed profiles, Blackstone River

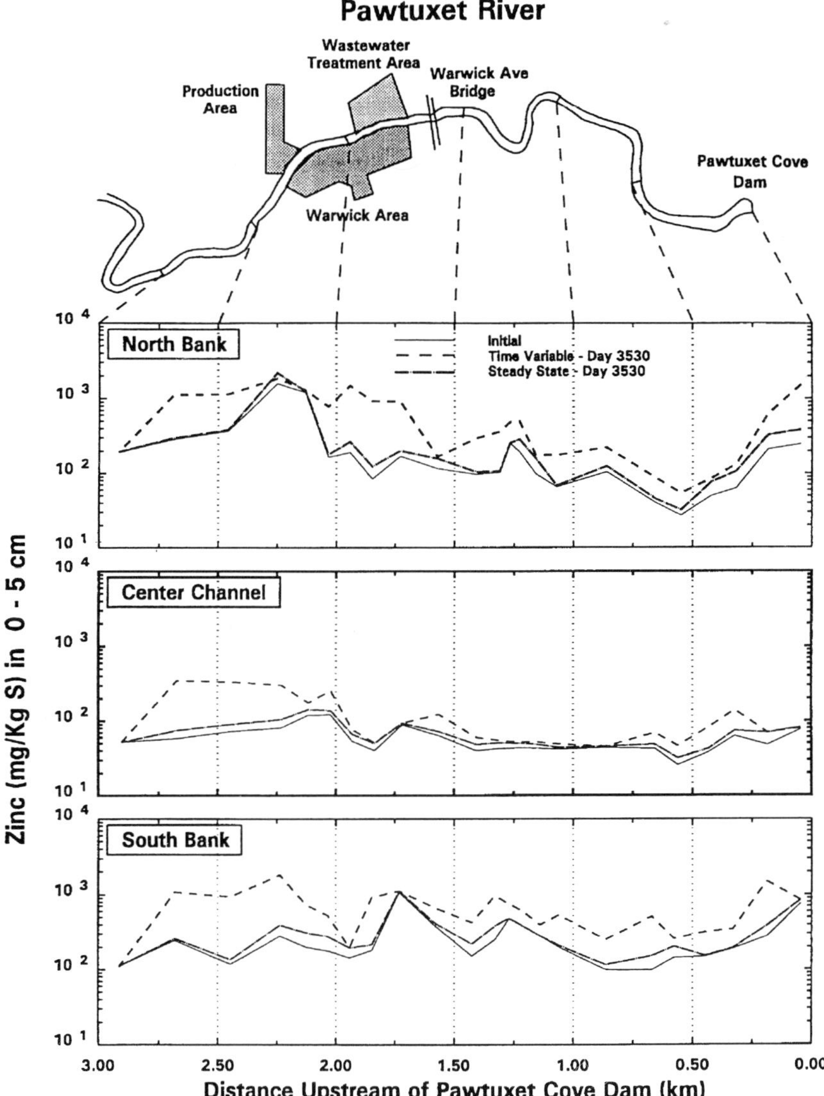

Figure 3 Zinc concentrations predicted in surficial sediments of the Pawtuxet River after 10 yrs (3520 d) of a constant loading at the upstream boundary. Time Variable = simulation of the variable hydrograph. Steady State = constant low-flow condition.

gests that the low-flow, steady-state models may not be useful for relating metals discharge and sediment contamination.

References

Di Toro DM, Mahoney JD, Hansen DJ, Berry WJ. A model of the oxidation of iron and cadmium sulfide in sediments. *Environ Toxicol Chem* (submitted).

HydroQual. 1982. Chemical quality analysis of the lower Genesee River. Final Report to Eastman Kodak Company, Rochester NY.

HydroQual. 1995b. Development of total maximum daily loads and wasteload allocations for toxic metals in NY/NJ Harbor. Final Report to U.S. Environmental Protection Agency (USEPA), Region II.

HydroQual. 1995a. Contaminant transport and fate modeling of the Pawtuxet River, Rhode Island. Final Report to Ciba Corporation, Toms River NJ.

[LTI] Limno-Tech, Inc. 1993. Field application of a steady state mass balance model to heavy metals in the Blackstone River. Washington DC: Final Report to U.S. Environmental Protection Agency (USEPA) Office of Wastewater Enforcement and Compliance.

SETAC

A Professional Society for Environmental Scientists and Engineers and Related Disciplines Concerned with Environmental Quality

The Society of Environmental Toxicology and Chemistry (SETAC), with offices in North America and Europe, is a nonprofit, professional society that provides a forum for individuals and institutions engaged in the study of environmental problems, management and regulation of natural resources, education, research and development, and manufacturing and distribution.

Goals

- Promote research, education, and training in the environmental sciences
- Promote systematic application of all relevant scientific disciplines to the evaluation of chemical hazards
- Participate in scientific interpretation of issues concerned with hazard assessment and risk analysis
- Support development of ecologically acceptable practices and principles
- Provide a forum for communication among professionals in government, business, academia, and other segments of society involved in the use, protection, and management of our environment

Activities

- Annual meetings with study and workshop sessions, platform and poster papers, and achievement and merit awards
- Monthly scientific journal, *Environmental Toxicology and Chemistry*, SETAC newsletter, and special technical publications
- Funds for education and training through the SETAC Scholarship/Fellowship Program
- Chapter forums for the presentation of scientific data and for the interchange and study of information about local concerns
- Advice and counsel to technical and nontechnical persons through a number of standing and *ad hoc* committees

Membership

SETAC's growing membership includes more than 5,000 individuals from government, academia, business, and public-interest groups with technical backgrounds in chemistry, toxicology, biology, ecology, atmospheric sciences, health sciences, earth sciences, and engineering.

If you have training in these or related disciplines and are engaged in the study, use, or management of environmental resources, SETAC can fulfill your professional affiliation needs. Membership categories include Associate, Student, Senior Active, and Emeritus.

For more information, contact SETAC, 1010 North 12th Avenue, Pensacola, Florida, T 904 469 1500, F 904 469 9778, E setac@setac.org.